Learning Java Bindings for OpenGL (JOGL)

By

Gene Davis

1663 LIBERTY DRIVE, SUITE 200
BLOOMINGTON, INDIANA 47403
(800) 839-8640
WWW.AUTHORHOUSE.COM

All products mentioned herein are trademarks or registered trademarks of their respective owners.

© 2004 Gene Davis.
All Rights Reserved.

A permission is granted for readers to use and distribute any source code in this book. All other rights are reserved.

First published by AuthorHouse 10/19/04

ISBN: 1-4208-0362-X (sc)

Library of Congress Control Number: 2004098367

Printed in the United States of America
Bloomington, Indiana

This book is printed on acid-free paper.

THE SOFTWARE IN THIS BOOK AND THE RELATED INTERNET SITE IS PROVIDED BY THE COPYRIGHT HOLDER "AS IS" AND ANY EXPRESS OR IMPLIED WARRANTIES, INCLUDING, BUT NOT LIMITED TO, THE IMPLIED WARRANTIES OF MERCHANTABILITY AND FITNESS FOR A PARTICULAR PURPOSE ARE DISCLAIMED. IN NO EVENT SHALL GENE DAVIS, OR GENE DAVIS SOFTWARE BE LIABLE FOR ANY DIRECT, INDIRECT, INCIDENTAL, SPECIAL, EXEMPLARY, OR CONSEQUENTIAL DAMAGES (INCLUDING, BUT NOT LIMITED TO, PROCUREMENT OF SUBSTITUTE GOODS OR SERVICES; LOSS OF USE, DATA, OR PROFITS; OR BUSINESS INTERRUPTION) HOWEVER CAUSED AND ON ANY THEORY OF LIABILITY, WHETHER IN CONTRACT, STRICT LIABILITY, OR TORT (INCLUDING NEGLIGENCE OR OTHERWISE) ARISING IN ANY WAY OUT OF THE USE OF THIS SOFTWARE, EVEN IF ADVISED OF THE POSSIBILITY OF SUCH DAMAGE.

"Rosebud"

For questions, comments or corrections email the author at:
support@genedavissoftware.com
Please mention the name of the book in the subject line.

Table of Contents

Preface viii

Author's Note x

Acknowledgments x

Chapter One Hello JOGL 1

Chapter Two Using What We've Seen 19

Chapter Three Animations 53

Chapter Four Using Events with GLCanvas 70

Chapter Five Going 3D 95

Chapter Six Drawing Geometric Primitives 107

Chapter Seven First Person Movement in 3D Space 117

Chapter Eight Lights 126

Chapter Nine Textures 136

Appendix A JOGL Online Resources 153

Appendix B OpenGL Online Resources 155

Appendix C Matrix Math 157

Preface

JOGL is likely the biggest improvement in Java since the introduction of Swing. Many people have pointed to the lack of OpenGL in Java as one of its greatest weaknesses. The introduction of Java Bindings for OpenGL opens new vistas for Java in fields that require fast powerful 3D models. OpenGL is used where ever 3D illustrations are useful. This includes games, medical, applications and architecture.

This book is for the beginner. When I say beginner, I mean anyone who has just learned some Java and wants to do something even cooler in Java. JOGL is not hard once you understand it, but the learning curve can be nasty. This book only gives basics that will make the learning curve more manageable. Hopefully there will be move advanced manuals available by the time you finish reading it. If not, maybe I'll write one.

No OpenGL or C experience is expected, and only minimal Java experience is expected.

If you are well-grounded in advance Java APIs, OpenGL (the C variety) and math, then this book will be an easy read. You should be able to learn the basics quickly and even skim some portions that are just a review to you. If you are just starting to learn to program then you may need to refer to other manuals occasionally for information you may not have learned yet.

My best advice to beginners is, "Don't be discouraged. We were all beginners at one time."

I go into some detail on math where it seems appropriate. If you want to program anything significant with the JOGL APIs, expect to learn some more advanced math. You don't have to learn it all at once though.

I have given many stand-alone examples in the course of this book. Code examples could have been simplified if they weren't stand-alone, but

I know how annoying it can be to be reading a chapter of a book and not be able to complete a code example, because you didn't read some earlier chapter.

As for the code examples, be sure to type them in. This is important to do for practice so that you can learn to write your own JOGL based programs. Cut and paste won't teach you what you need to learn.

You can check out my website for additional resources not included in the book. It is http://www.genedavissoftware.com. I hope you learn to love JOGL as much as I have!

Author's Note

Computer code doesn't always look good in print, nor can you cut and paste from a book. Look for the code examples from this book, corrections, precompiled JARs and additional help on the author's website:

http://www.genedavissoftware.com/

Acknowledgments

Special thanks should be given to the following for their input: Craig Mattocks, for the improvements on the SecondJoglApp.java; Stephyn Butcher for suggested improvements; Blake Bishop; Holly Bishop; and the developers who have worked so hard to make JOGL a reality.

I'd also like to thank my wife for her support, and don't want to leave out my parents, my children, my brother, my grandparents, Santa Claus, the Great Pumpkin (surely you remember him) or the Easter Bunny.

Thanks everyone!
Gene Davis
www.genedavissoftware.com

Chapter One
Hello JOGL

First There Was OpenGL

For some years now, a programmer that wanted to create a graphics intensive program that could be sold to users of different Operating Systems had one choice -- OpenGL. The GL stands for Graphics Library. OpenGL is a registered trademark of SGI. OpenGL manifests itself as a cross platform C programming API. In reality though, it is a hardware-independent specification for a programming interface.

Most of the reasons that people have given me over the years for the lack of a future for Java, especially in the gaming industry, was that you can't use OpenGL in Java. They rightly pointed out that the fastest 2D and 3D applications use OpenGL.

OpenGL is for making graphics. It is fast. Most of the time it is hardware accelerated. It seems that OpenGL can do anything visually that you would want to do.

Unfortunately OpenGL is written for C. Let's face it, C is not the most popular language for programming complex applications. One of the biggest drawbacks to OpenGL is that you can't make it do anything without a window to put your graphics in, but OpenGL doesn't provide a means for you to create windows. This makes OpenGL hard to learn for beginners.

Luckily GLUT was introduced and made dealing with windows, buttons and events generated by users easier to add to OpenGL heavy applications. Still learning OpenGL in C or even C++ can be painful for new programmers or programmers that want to use true Object Oriented Programming.

Then Came JOGL

Java is possibly the most popular true Object Oriented Programming language. There have been many attempts to marry OpenGL with Java, but the first one that made everyone stand up and take notice was JOGL. The reason for this is that this effort is supported by Sun (the creators of Java) and SGI (the creators of OpentGL).

Nowadays JOGL is developed by Game Technlogy Group at Sun Microsystems. It started out life as Jungle developed by Ken Russel and Chris Kline. Russell is a Sun Microsystems employee working on the HotSpot Virtual Machine with many years of 3D experience. Kline works for Irrational Games and also is very experienced with 3D graphics.

I am personally grateful for their efforts and the efforts of all those who are working on JOGL. There have been several attempts at providing access to OpenGL through a friendly Java API. Among these have been Java 3D, OpenGL for Java Technology (gl4java) and Lightweight Java Game Library (LWJGL). JOGL is the first that I felt comfortable with.

JOGL is the Sun supported set of Java class bindings for OpenGL. Wow! That was a mouthful.

OpenGL is used to display 3D Models. It is powerful, fast and perhaps the greatest thing to happen to Java since Swing was introduced. Using OpenGL through JOGL, you will be able to make cool games, or model situations that could be too expensive to create.

Thick tomes have been written describing OpenGL. They will be useful once you know your way around, but not yet. You need to learn how this all applies to the Java APIs that expose OpenGL to you. You also need some basic introductions to net.java.games.jogl.*, and perhaps some refreshers on math.

Got JOGL?

If you want to use JOGL, you will need to get 'jogl.jar' and its accompanying native code. I dream of the day it is standard with the Java installation, but for now that is just a well placed dream.

The first trick is finding the binaries for your OS and extracting them. I found them at https://games-binaries.dev.java.net/build/index.html. Every OS is different but there are two parts to installing. The jogl.jar must be placed in the system classpath and the binary library must be placed wherever libraries go in your OS. If you're lucky you will have an installer to do it for you. If you don't have an installer and don't know where to look for information on placing everything on your computer, you can start with the links I've provided in Appendix A. Our first code example will be written specifically to test whether you've installed everything correctly, so you don't need to stress about testing your installation until then.

Javadocs for JOGL

The Javadocs may be obtained at the same location as the binary distribution of JOGL. That is https://games-binaries.dev.java.net/build/index.html. The Javadocs will be named something similar to jogl-1.0-usrdoc.tar.

If you browse the net.java.games.jogl package, you'll quickly notice that some of the classes are huge. GL is a perfect example of this. Don't be put off by this. You'll find out quickly that you're able to do some pretty sophisticated work even with just a small amount of JOGL knowledge.

The classes you might want to glance at now are:
GLDrawable
GLCanvas
GLJPanel
GLCapabilities
GLDrawableFactory

These will be you're basic interface into the world of graphics. If you remember, earlier I mentioned that one of the greatest drawbacks for beginners learning OpenGL was the lack of a windowing system standard to learn on. GLUT helps a long ways in that regard for our C counterparts, but we have Swing and the AWT. It's very likely you have already used AWT or Swing, so are not going to feel like you are learning everything from scratch.

This is a good thing. After a very brief introduction to getting a Component for JOGL up on the screen, we won't need much work to have you running pretty cool and hip apps!

Review of Java Concepts

I do assume that you know Java, but let's go through a quick review of some of the more advanced topics. You can find more detailed and lengthy descriptions in books devoted to teaching the Java language and APIs.

Gene Davis

Floats and Doubles

One of the things I rarely see discussed properly in Java text books is how to use the primitives float and double. Even long primitives are not really given the respect they deserve, but we'll stick to floats and doubles here.

In Java, how do you assign a float? If you try an assignment like this,

```
float f = 1.0;
```

you will quickly get an error. Why? Because '1.0', like every decimal number is assumed to be a double unless cast or specified otherwise. You will get a warning about possible loss of precision from the compiler.

You could do an assignment like

```
float f = (float) 1.0;
```

but why would you want to do a cast every time you type a float number? Instead, you will want to do an assignment such as

```
float f = 1.0f;
```

The 'f' at the end of the number tells the compiler that this is a float not a double. Remember that there is no space between the '0' and the 'f' in the example above.

Here's another gotcha that you will likely run into when dealing with decimals in Java or other computer languages. The math is approximated, not exact. For instance, this program would be expected by many to print "d1 == d2" instead on my system it prints out "1.0000000128746032 != 1.0000000238418578".

```
public class Example {
  public static void main (String[] args) {
    float f1 = 0.999f;
    float f2 = 1.1f;

    double d1 = f1 + .001;
    double d2 = f2 - .1;

    if (d1 == d2) System.out.println("d1 == d2");
    else System.out.println(d1+" != "+d2);
  }
}
```

So '1.0' does not always equal itself. The conversion to doubles is what gets us confused in this example. It is close though and that is likely to be good enough if you keep in mind that decimal math won't always give you the exact results you would get with a calculator. When precision is needed, stick with ints or longs and convert to floats and doubles later.

Event-Listener Model

As you learned to use Swing or the AWT, you were introduced early on to the Event-Listener Model. This is the model that Java uses for responding to the user's actions. It is often used for thread communication within a program. Sometimes events are not generated by the user at all.

The first part of the event-listener model is the event object. The event represents some action that has occurred because of the user or a thread that just got lonely and wanted to chat. Okay, threads don't get lonely, but you get the idea.

The second part of the model is the listener. Ever been give a quarter to call someone who cares? Well, the listener cares. He cares so much that he implemented a needed listener interface and got registered as a caring listener.

Third is the generator. We register listeners with event generators. The generator calls any listeners who cared enough to register themselves (usually through addSuchAndSuchListener() calls.) It is the event generator that actually creates the event, though the trigger that creates an event probably comes from a thread's action (such as a while loop) or the user's action (such as a mouse click).

GLEventListener

One listener that you must use in JOGL is the GLEventListener. Whenever your JOGL program is ready to draw some OpenGL, it will look and see if there are any GLEventListeners that were added.

After looking up the GLEventListener, the listener is called in one of its four methods and this determines what is drawn using OpenGL. A GLEventListener must implement display(), displayChanged(), init() and reshape(). You will find that display() and init() are extremely useful, but the other two may not even need to be used. In other words, focus on learning to implement init() and display().

You will see your first GLEventListener shortly.

Implements v.s. Extends

'Implementing' and 'extending' both are used to make your class into something more than itself.

Implementations 'implement' interfaces. Interfaces don't do anything. Really they don't. All they are is a pattern to follow in making a class. If the pattern is not followed accurately, the compiler stops you and tells you to follow the pattern.

Interfaces may be extended by interfaces or implemented by classes. Classes may be extended by classes. Implementing means that you made the interface's methods work. Extending means you get what you get, though you may improve on what you're given.

Okay, that was brief. It was a review, not a tutorial. If you haven't ever been introduced to 'implements' or 'extends', you're going to need to pull up a tutorial and learn them and it would be a good idea to get a grip on polymorphism too.

Polymorphism

Polymorphism is when a class implements an interface or extends a class and it is treated like the class or interface it is like, rather than itself.

If you think of the Harry Potter movies or books, Professor Snape hates Harry Potter. If you read into it, he really hates Harry's father. So Professor Snape is using polymorphism to take out his hatred on Harry's father even though the object he is treating so badly is really the son of Mr. Potter -- that is Harry, not Harry's father. Harry is being treated like someone he is not, but someone he shares many traits with (such as looks).

Polymorphism is valuable to programmers, because they often know what they wish to do with a class that hasn't been written yet. So they make an interface and demand that the class or classes that they use in the future implement that interface. That way they may use any of the classes that they or anyone else write without changing their code.

Mixing Heavy and Light Components

The AWT uses heavy components. Swing uses light components. Many programmers will shiver at the mention of mixing the too. No one should worry, just learn the rules and you will likely be fine. The main rule is that heavy components always look like they are on top even if they shouldn't look like they are, unless they own the light component or the

Learning Java Bindings for OpenGL (JOGL)

light component is shielded from the heavy component by its owner who is a heavy component.

That sounded much worse than it is, and don't be too hard on yourself if you stick to not mixing AWT and Swing components. However, we will be mixing them in this book. Most of the programs have been tested at least in two implementations of the JVM and have not shown any issues for concern.

GLDrawable, GLCanvas and GLJPanel

GLDrawable is an interface. All of JOGL's OpenGL drawing will happen in GLDrawables. GLCanvas and GLJPanel both are classes that implement GLDrawable. As far as your program is concerned all GLCanvases and all GLJPanels are GLDrawables. This is polymorphism. The GLDrawables being passed around in the GLEventListeners that you implement will really be GLCanvases and GLJPanels, but you won't care what they really are because you'll be using polymorphism to treat them like GLDrawables.

You will benefit from a glance at the GLDrawable interface in the javadocs. The main methods that you will use are addEventListener(), getGL() and getGLU().

If you are already familiar with OpenGL in its more tradition C manifestation, then GL and GLU should sound familiar. They are the two main headers containing definitions used in OpenGL.

It is worth mentioning that the GLCanvas is an AWT component. It is heavy, but faster at rendering OpenGL commands than GLJPanel currently is. GLJPanel is a Swing component, and hopefully catches up to GLCanvas for speed someday. GLJPanel is a light component.

JNI

You won't need to know any JNI to use JOGL. You will benefit from a basic understanding of it. I'm not going to say much here, because it will scare beginners. JNI is used to access C libraries that would be faster than methods written in Java. OpenGL is one such library. It is one of the best graphics libraries and available on most modern computers. JOGL uses JNI to access OpenGL libraries for graphics.

Gene Davis

GlueGen ... Almost As Cool As JOGL?

As you should be aware, OpenGL is written for C programmers. This means that for Java to take advantage of it there has to be some native interface. This means JNI, which isn't fun or pretty, must be written to make this connection.

OpenGL is pretty big. Writing all those connections takes time. To make things just a little more difficult there are plenty of vendor specific features and OpenGL keeps improving, which means there are changes to keep up with. In short, it has been pretty hard for the "anyone" trying to keep up with OpenGL to write a Java to native interface that is all encompassing.

Enter the JOGL folks. They decided to take advantage of the C header files and write some code that would do all the JNI work for them. They called it GlueGen. GlueGen parses the C header files and then magically creates the needed Java and JNI code necessary to connect to those native libraries. This means that updates to OpenGL can be added quickly to JOGL.

Very cool is an understatement. As Chris Kline put it, "IMHO, the 'GlueGen' generator that Jogl uses to generate the GL binding is more valuable than the binding itself." (http://www.puppygames.net/forums/viewtopic.php?t=38&start=30).

Hello World!

I'm a firm believer in tradition, so of course we will start with a "Hello World". This Hello World will examine our installation and tell us whether all or part is installed correctly. Remember there are two parts to the JOGL installation. There is the Java library in a Jar file and the native code in another library.

Here is our program:

```
import net.java.games.jogl.*;

public class HelloWorld {
  public static void main (String args[]) {
    try {
      System.loadLibrary("jogl");
      System.out.println(
          "Hello World! (The native libraries are installed.)"
      );
```

Learning Java Bindings for OpenGL (JOGL)

```
      GLCapabilities caps = new GLCapabilities();
      System.out.println(
        "Hello JOGL! (The jar appears to be
available.)"
      );
    } catch (Exception e) {
      System.out.println(e);
    }
  }
}
```

First this program tests to see if the native and java libraries are installed correctly. JOGL is installed properly only when the 'jogl.jar' and the native library, named something like 'libjogl.jnilib' or 'jogl.dll', are both installed. If the native library is not accessible, this program will throw a "java.lang.UnsatisfiedLinkError" Exception. If the jar is not installed in the classpath, then program will not even compile. The javac compiler will say something similar to "package net.java.games.jogl does not exist". When this class compiles and runs without exceptions, you are ready to continue with learning JOGL.

A Good Template

Let's move on to a couple of classes that you may find useful to use as a template while messing around with JOGL. I've used them as templates more than once. Feel free to use them however you like.

This template is made up of two classes. The first is SimpleJoglApp shown below, and the second is SimpleGLEventListener shown after a brief description. You will need to type both in to compile the template.

The main app....

```
import java.awt.*;
import java.awt.event.*;
import javax.swing.*;
import net.java.games.jogl.*;

/**
 * This is a basic JOGL app. Feel free to
 * reuse this code or modify it.
 */
public class SimpleJoglApp extends JFrame {

  public static void main(String[] args) {
```

```java
    final SimpleJoglApp app = new SimpleJoglApp();

    // show what we've done
    SwingUtilities.invokeLater (
      new Runnable() {
        public void run() {
          app.setVisible(true);
        }
      }
    );
}

public SimpleJoglApp() {
    //set the JFrame title
    super("Simple JOGL Application");

    //kill the process when the JFrame is closed
    setDefaultCloseOperation(JFrame.EXIT_ON_CLOSE);

    //only three JOGL lines of code ... and here they are
    GLCapabilities glcaps = new GLCapabilities();
    GLCanvas glcanvas =
        GLDrawableFactory.getFactory().createGLCanvas(glcaps);
    glcanvas.addGLEventListener(new SimpleGLEventListener());

    //add the GLCanvas just like we would any Component
    getContentPane().add(glcanvas, BorderLayout.CENTER);
    setSize(500, 300);

    //center the JFrame on the screen
    centerWindow(this);
}

public void centerWindow(Component frame) {
    Dimension screenSize =
        Toolkit.getDefaultToolkit().getScreenSize();
    Dimension frameSize  = frame.getSize();

    if (frameSize.width  > screenSize.width )
        frameSize.width  = screenSize.width;
```

Learning Java Bindings for OpenGL (JOGL)

```
    if (frameSize.height > screenSize.height)
      frameSize.height = screenSize.height;

    frame.setLocation (
      (screenSize.width  - frameSize.width ) >> 1,
      (screenSize.height - frameSize.height) >> 1
    );
  }
}
```

That was it. Let's focus on the three lines of JOGL specific code in this first class. To start:

```
GLCapabilities glcaps = new GLCapabilities();
```

This determines what OpenGL/graphics features are available to our JOGL libraries and the JVM.

Next:

```
GLCanvas glcanvas =
    GLDrawableFactory.getFactory().createGLCanvas(glcaps);
```

We cannot create GLCanvases or GLJPanels. We need to have them created for us by a GLDrawableFactory. So, we retrieve a GLDrawableFactory using GLDrawableFactory's static method, getFactory().

Now we have a GLDrawableFactory, so we use its createGLCanvas() method to create a GLCanvas to draw on. We could have used the createGLJPanel() method instead if we had wanted a Swing component instead of an AWT component.

Notice that we passed in the GLCapabilities object we created earlier. This allows the GLDrawable to be created properly.

Finally we are ready to add a GLEventListener to the GLCanvas.

```
glcanvas.addGLEventListener(new SimpleGLEventListener());
```

Our implementation of GLEventListener is SimpleGLEventListener. It will take care of any drawing that needs to be done when it receives a call from the GLDrawable our one and only GLCanvas. As you will see, we decide not to draw anything in this program.

Remember, if I call a class by its parent's name, that is okay because of polymorphism. Now for the GLEventListener....

```java
import java.awt.*;
import java.awt.event.*;
import net.java.games.jogl.*;

/**
 * For our purposes only two of the
 * GLEventListeners matter. Those would
 * be init() and display().
 */
public class SimpleGLEventListener implements
GLEventListener {

  /**
   * Take care of initialization here.
   */
  public void init(GLDrawable drawable) {

  }

  /**
   * Take care of drawing here.
   */
  public void display(GLDrawable drawable) {

  }

  /**
   * Called when the GLDrawable (GLCanvas
   * or GLJPanel) has changed in size. We
   * won't need this, but you may eventually
   * need it -- just not yet.
   */
  public void reshape(
                      GLDrawable drawable,
                      int x,
                      int y,
                      int width,
                      int height
                     ) {}

  /**
   * If the display depth is changed while the
```

```
 * program is running this method is called.
 * Nowadays this doesn't happen much, unless
 * a programmer has his program do it.
 */
public void displayChanged(
                            GLDrawable drawable,
                            boolean modeChanged,
                            boolean deviceChanged
                          ) {}
}
```

That is the heart of the JOGL work we will do. Notice the UML graphic below. SimpleJoglApp is a JFrame. It contains our GLDrawable which is actually a GLCanvas, but don't tell him that. We add() the SimpleGLEventListener which implements GLEventListener to the GLCanvas so the GLCanvas knows we care if he wants any OpenGL work done. GLDrawables can talk your ear off, so you'll want to make sure your GLEventListener is optimized, ... for real.

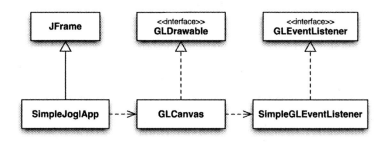

This app may look a bit scrambled depending on your OS. This is to be expected, because you are just displaying random bits of memory at this point. So congradulations on your new found graphics talents.

Gene Davis

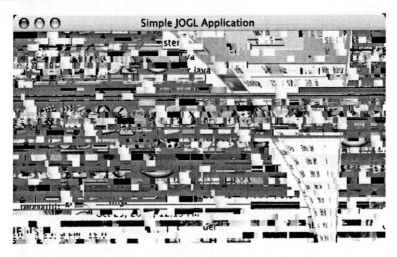

You're Ready For the Real Thing

After you've familiarized yourself with the previous example, let's make a pretty picture.

Here is your next app. Make sure you type this and all examples in. Debugging and messing around with them will serve to quickly teach you how they work.

```
import java.awt.*;
import java.awt.event.*;
import javax.swing.*;
import net.java.games.jogl.*;

/**
 * This is a basic JOGL app. Feel free to
 * reuse this code or modify it.
 */
public class SecondJoglApp extends JFrame {

  public static void main(String[] args) {
    final SecondJoglApp app = new SecondJoglApp();

    //show what we've done
    SwingUtilities.invokeLater (
      new Runnable() {
        public void run() {
          app.setVisible(true);
        }
```

```java
      }
    );
  }

  public SecondJoglApp() {
    //set the JFrame title
    super("Second JOGL Application");

    //kill the process when the JFrame is closed
    setDefaultCloseOperation(JFrame.EXIT_ON_CLOSE);

    //only three JOGL lines of code ... and here they are
    GLCapabilities glcaps = new GLCapabilities();
    GLCanvas glcanvas =
        GLDrawableFactory.getFactory().createGLCanvas(glcaps);
    glcanvas.addGLEventListener(new SecondGLEventListener());

    //add the GLCanvas just like we would any Component
    getContentPane().add(glcanvas, BorderLayout.CENTER);
    setSize(500, 300);

    //center the JFrame on the screen
    centerWindow(this);
  }
  public void centerWindow(Component frame) {
    Dimension screenSize =
        Toolkit.getDefaultToolkit().getScreenSize();
    Dimension frameSize = frame.getSize();

    if (frameSize.width  > screenSize.width )
        frameSize.width  = screenSize.width;
    if (frameSize.height > screenSize.height)
        frameSize.height = screenSize.height;

    frame.setLocation (
      (screenSize.width  - frameSize.width ) >> 1,
      (screenSize.height - frameSize.height) >> 1
    );
  }
```

}

Notice hardly anything changed in the first class. Only changes that affected the names of the class, frame and GLEventListener were made. Hopefully you've read the included comments in the code, or you're going to miss out on the explanation of what is going on.

The GLEventListener we've implemented does have some changes to allow us to draw something nicer than in the last example.

```
import net.java.games.jogl.*;

/**
 * For our purposes only two of the GLEventListeners matter.
 * Those would be init() and display().
 */
public class SecondGLEventListener implements GLEventListener {

  /**
   * Take care of initialization here.
   */
  public void init(GLDrawable gld) {
    GL gl = gld.getGL();
    GLU glu = gld.getGLU();

    gl.glClearColor(0.0f, 0.0f, 0.0f, 1.0f);

    gl.glViewport(0, 0, 500, 300);
    gl.glMatrixMode(GL.GL_PROJECTION);
    gl.glLoadIdentity();
    glu.gluOrtho2D(0.0, 500.0, 0.0, 300.0);
  }

  /**
   * Take care of drawing here.
   */
  public void display(GLDrawable drawable) {

    float red = 0.0f;
    float green = 0.0f;
    float blue = 0.0f;

    GL gl = drawable.getGL();
```

```
    gl.glClear(GL.GL_COLOR_BUFFER_BIT);

    gl.glPointSize(5.0f);

    for (int i=0; i<50; i++) {

      red   -= .09f;
      green -= .12f;
      blue  -= .15f;

      if (red   < 0.15) red   = 1.0f;
      if (green < 0.15) green = 1.0f;
      if (blue  < 0.15) blue  = 1.0f;

      gl.glColor3f(red, green, blue);

      gl.glBegin(GL.GL_POINTS);
          gl.glVertex2i((i*10), 150);
      gl.glEnd();

    }
  }

  public void reshape(
                      GLDrawable drawable,
                      int x,
                      int y,
                      int width,
                      int height
                     ) {}
  public void displayChanged(
                      GLDrawable drawable,
                      boolean modeChanged,
                      boolean deviceChanged
                     ) {}
}
```

That's our first interesting JOGL program. Below is the output. It has plenty of pretty colors.

Gene Davis

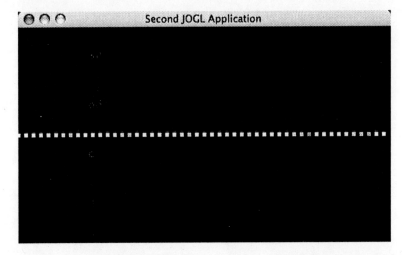

If you look at the implementation of the GLEventListener you may feel a bit overwhelmed. If you are experienced with OpenGL using C, you can probably divine what is going on. Don't worry if you find it confusing and are afraid I'm going to ask you to commit it to memory. I will. Just not yet.

The rest of the book will explain what it happening in SecondGLEventListener in this example. For now try guessing. Mess around with the code. Try making two lines or a line that is diagonal instead of horizontal, or make all the dots blue or red. Have some fun. That's how you're going to learn JOGL after all.

Chapter Two
Using What
We've Seen

Coordinate Systems

There is no sense in waiting until you know the best way to do things before using OpenGL. Sure you're going to find out a few months later that there was a better way to do this or that, but you'll never get to that point if you don't use what you've learned. We're going to take that principle to heart and start off showing you that you've already learned a lot.

Coordinate systems may be something that you've banished to the darker regions of your memory along with memories of Junior High School. It's time to dust off those memories.

Do you remember the Cartesian coordinate system? It consists of two perpendicular axes [lines]. One is the y axis and the other is the x axis.

Here is a typical coordinate system with a point mapped out on it.

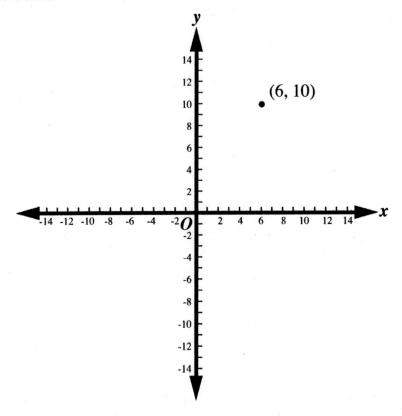

You'll notice that there are numbers all over the place. Usually, they're not all written out, but they're always there. All points on the coordinate system correspond to distances along the y axis or along the x axis away from the origin. The origin is where the two axes meet. We've marked the origin with an O.

So we've indicated the position of one point on the graph. It is (6, 10). The 6 corresponds to 6 on the x axis away from the origin. The 10 indicates the point is 10 from the origin on the y axis.

Notice some positions are negative and some are positive. It would get confusing if 2 left of the origin looked like 2 right of the origin when we mapped out points. If there wasn't any negatives how would we be able to tell where the point (1, 2) was? It could be in four different places, however if negatives are used (-1, 2) is very easy to spot. Can you find it?

If you're lost at this point, your in trouble. Find a good introduction to geometry or introduction to graphs and functions book and plan on reading some of it. I'd recommend checking out a college book store, or maybe a big general book store. If money's an issue, check out the internet for tutorials. There are plenty out there.

You can continue reading either way, but plan on spending time on learning about graphs.

During this chapter we will be using the coordinate system extensively. We will also cover drawing lines and the old Trig. standbys -- Sine, Cosine and Tangent. Eventually in your pursuit of graphics and JOGL, you will want to become thoroughly familiar with these, but that is then and now is now. I'll explain briefly anything that matters, and you can always use this book for reference.

glViewport

You may remember seeing a line of code similar to:

```
gl.glViewport(0, 0, 500, 300);
```

This is using the GL Object to set the area of the screen to be seen. This is a slight simplification, but will be a good enough definition for our purposes. So remember, glViewport() clips the area to be viewed.

Imagine you have a big piece of paper. Now imagine that you cut a rectangle out of it. If you placed the paper over your monitor, you would only see the portion that wasn't covered by the paper. In other words, you would see through the rectangular hole in the paper and that's all.

The rectangular hole in the paper acts the same as the glViewport().

gluOrtho2D

The gluOrtho2D() method is used to set the coordinate system that appears in your GLCanvas. This is sort of like the numbers we have on the Cartesian coordinate system in the first section of this chapter. You can choose any numbers that you find useful for what you are doing. For now you will set the glViewport() and the gluOrtho2D() to the same values. Remember, the gluOrtho2D() sets the coordinates the GLCanvas maps to.

glClearColor, GL_COLOR_BUFFER_BIT and glClear

When you erase a black board what color does it become? The correct answer is black, but I'm sure someone out there is thinking, "Well that depend on the color of chalk you used." To that someone, I'm not talking to you anymore.

Gene Davis

When you clean a white board what color does it become? I'm sure everyone said white, since we're not talking to the naysayers anymore.

The GLCanvas that we're doing our drawing on is a color too. What color is it when we clear it? I just know someone said, "Canvas color." You there, ... Go join the naysayers! Canvas color indeed....

The truth is that you, the programmer choose the color. The method that chooses the color looks like this:

```
glClearColor(0.0f, 0.0f, 0.0f, 1.0f)
```

The first three floating point numbers represent red, green and blue. The forth number is for the alpha layer. If you know what that is, great. Any graphic designers reading this, probably have dealt with alpha layers before. I'll spare the rest of you the details for now.

The color of the cleared GLCanvas is chosen by mixing the red, green and blue elements using numbers between 0.0 and 1.0. Red would be:

```
glClearColor(1.0f, 0.0f, 0.0f, 1.0f)
```

Blue would be:

```
glClearColor(0.0f, 0.0f, 1.0f, 1.0f)
```

Green would be:

```
glClearColor(0.0f, 1.0f, 0.0f, 1.0f)
```

A valid question is how do I choose some other color? Find a simple painting program, or any program that lets you choose an RGB color. PhotoShop displays a nice color picker. Pick a color that you like. Let's choose purple.

RGB colors usually assign the values for red, green and blue to numbers between 0 and 255. Purple, in this case, is assigned the values of 124, 11 and 183, for red, green and blue respectively. These are hardly values between 0.0 and 1.0.

No problem. If you're a little fuzzy on your math, here's what you need to do. Take each of the numbers and divide by 255. Now you have a number between 0.0 and 1.0. In this case the numbers are:

```
glClearColor(0.486f, 0.043f, 0.718f, 1.0f)
```

Math is your friend. It's also not too boring when it is being useful.

Now that you've chosen a color, you can clear the GLCanvas to your color any time you want. Just use the GL command:

```
glClear(GL.GL_COLOR_BUFFER_BIT)
```

So what color what color is a GLCanvas when it's cleared? Canvas color of course.

glColor3f

It's time to choose your own colors. You need to choose a color to draw in it. Remember that you can use any RGB color picker to pick a pretty color to draw in, and then convert the base 256 numbers for red, green and blue to the proper float value by dividing it by 255.

To set your current drawing color, use your GL object to call the command glColor3f(). It looks something like this:

```
gl.glColor3f(.9f, .12f, .15f);
```

Personally, I always like to predefine my red, green and blue values in float variables, so the actual line looks like this:

```
gl.glColor3f(red, green, blue);
```

That just seems to be easier on the eyes, and trust me you will want to make anything you can easy on your eyes when you have a 30,000 line Java program in front of you.

Making a graph

Now let's apply some of the knowledge we've gained. We're going to make a Cartesian coordinate system similar to the one we showed earlier in the chapter. Don't stress. It's actually not much work.

Here's the code for the main class:

```
import java.awt.*;
import java.awt.event.*;
import javax.swing.*;
import net.java.games.jogl.*;

/**
```

Gene Davis

```
 * This is a basic JOGL app. Feel free to
 * reuse this code or modify it.
 */
public class OurGraphApp extends JFrame {

    public static void main(String[] args) {
        final OurGraphApp app = new OurGraphApp();

        // show what we've done
        SwingUtilities.invokeLater (
            new Runnable() {
                public void run() {
                    app.setVisible(true);
                }
            }
        );
    }

    public OurGraphApp() {
        //set the JFrame title
        super("The Cartesian Coordinate System");

        //kill the process when the JFrame is closed
        setDefaultCloseOperation(JFrame.EXIT_ON_CLOSE);

        //only three JOGL lines of code ... and here they are
        GLCapabilities glcaps = new GLCapabilities();

        GLCanvas glcanvas =
        GLDrawableFactory.getFactory().createGLCanvas(glcaps);

        glcanvas.addGLEventListener(
            new OurGraphGLEventListener()
        );

        //add the GLCanvas just like we would any Component
        getContentPane().add(glcanvas, BorderLayout.CENTER);
        setSize(500, 300);

        //center the JFrame on the screen
```

```
            centerWindow(this);
    }

    public void centerWindow(Component frame) {
        Dimension screenSize =
            Toolkit.getDefaultToolkit().getScreenSize(
);
        Dimension frameSize  = frame.getSize();

        if (frameSize.width  > screenSize.width )
            frameSize.width  = screenSize.width;
        if (frameSize.height > screenSize.height)
            frameSize.height = screenSize.height;

        frame.setLocation (
            (screenSize.width  - frameSize.width ) >>
1,
            (screenSize.height - frameSize.height) >>
1
        );
    }
}
```

This is the code for the GLEventListener:

```
import java.awt.*;
import java.awt.event.*;
import net.java.games.jogl.*;

/**
 * For our purposes only two of the GLEventListeners
matter. Those would be
 * init() and display().
 */
public class OurGraphGLEventListener implements
GLEventListener {

    /////////////////////////////////////////////////
//
    // GLEventListener implementation
    //

    /**
     * Take care of initialization here.
     */
```

```
    public void init(GLDrawable gld) {
        GL gl = gld.getGL();
          GLU glu = gld.getGLU();

        gl.glClearColor(0.0f, 0.0f, 0.0f, 0.0f);

        gl.glLineWidth(2.0f);

        gl.glViewport(-250, -150, 250, 150);
        gl.glMatrixMode(GL.GL_PROJECTION);
        gl.glLoadIdentity();
        glu.gluOrtho2D(-250.0, 250.0, -150.0, 150.0);
    }

    /**
     * Take care of drawing here.
     */
    public void display(GLDrawable gld) {
        drawGraph( gld.getGL() );
    }


    public void reshape(
                    GLDrawable drawable,
                    int x,
                    int y,
                    int width,
                    int height
                ) {}
    public void displayChanged(
                        GLDrawable drawable,
                        boolean modeChanged,
                        boolean deviceChanged
                    ) {}

    //////////////////////////////////////////////
    //
    // Other methods
    //

    /**
     * In here we draw a Cartesian Coordinate System.
     */
    private void drawGraph(GL gl) {
```

```java
float red;
float green;
float blue;

gl.glClear(GL.GL_COLOR_BUFFER_BIT);

//////////////////////
//drawing the grid
red = 0.2f;
green = 0.2f;
blue = 0.2f;

gl.glColor3f(red, green, blue);

//You may notice I'm using GL_LINES here.
//Details of glBegin() will be discussed
   later.
gl.glBegin(GL.GL_LINES);

//draw the vertical lines
for (int x=-250; x<=250; x+=10) {
    gl.glVertex2d(x, -150);
    gl.glVertex2d(x, 150);
}

//draw the horizontal lines
for (int y=-150; y<=150; y+=10) {
    gl.glVertex2d(-250, y);
    gl.glVertex2d(250, y);
}

gl.glEnd();

/////////////////////////////
// draw the x-axis and y-axis
red = 0.0f;
green = 0.2f;
blue = 0.4f;

gl.glColor3f(red, green, blue);

gl.glBegin(GL.GL_LINES);

//line for y-axis
gl.glVertex2d(0, 140);
```

```
            gl.glVertex2d(0, -140);

            //line for x-axis
            gl.glVertex2d(240, 0);
            gl.glVertex2d(-240, 0);

            gl.glEnd();

            ////////////////////
            // draw arrow heads
            gl.glBegin(GL.GL_TRIANGLES);

            gl.glVertex2d( 0,  150);
            gl.glVertex2d(-5,  140);
            gl.glVertex2d( 5,  140);

            gl.glVertex2d( 0, -150);
            gl.glVertex2d(-5, -140);
            gl.glVertex2d( 5, -140);

            gl.glVertex2d(250, 0);
            gl.glVertex2d(240,-5);
            gl.glVertex2d(240, 5);

            gl.glVertex2d(-250, 0);
            gl.glVertex2d(-240,-5);
            gl.glVertex2d(-240, 5);

            gl.glEnd();

    }

}
```

OurGraphApp should look similar to this when run:

Learning Java Bindings for OpenGL (JOGL)

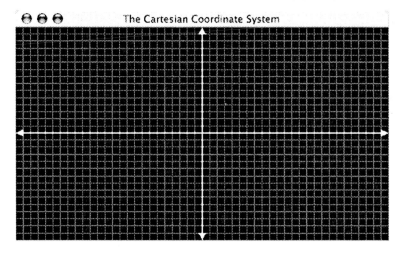

A description of what we've done is in order. We've created this program using two classes. They are the OurGraphGLEventListener and the OurGraphApp classes. OurGraphApp contains the main() method. Mostly we use the OurGraphApp class to start up the non JOGL specific code. The only JOGL specific code in OurGraphApp initializes the GLCapabilities class and retrieves a GLCanvas, then sets up the GLEventListener which does all the actual JOGL work.

OurGraphGLEventListener is a bit more JOGL intensive. It is our GLEventListener implementation. As I'm sure you recall, GLEventListeners draw pretty pictures using Java Bindings for OpenGL. The only two methods we care about for now are init() and display().

The init() method retrieves a copy of the current GL object and current GLU object. Many frugal programmers are in the habit of saving objects off, so they don't have to be retrieved later. Don't do this here. The GL and GLU objects may change while you're not looking, so get a new instance each time you can, just in case.

You should recognize some of the calls in the init() method. glClearColor(), glViewport() and gluOrtho2D() have been discussed earlier in this chapter.

Look at the glLineWidth(). We are drawing lines all over the place in the coordinate system. This is where we set its width.

Don't be fooled by the 2.0f for line width. This does not refer to the thickness of the line in the gluOrtho2D coordinate system. It refers to the thickness of the line in our monitor's resolution. Notice that if you make the window that the GLCanvas is in larger or smaller, the gluOrtho2D coordinate system is being stretched or shrunk, but the line widths remain the same.

In the display() method we just call the drawGraph() method. This will be useful later, because we will want to draw a graph in addition to what we are doing.

Look at the drawGraph() method. Examine where colors are set. Find the gl.glBegin(GL.GL_LINES) and the gl.glEnd() calls. Identify where the triangle arrow heads are drawn. We haven't discussed all of this yet, but you know enough now to spot where each part of the graph is drawn. You need to read through the code as well as type it in if you intend to learn JOGL.

Now mess with colors. Pick an RGB color of your own for the lines. Mess around with the glVertex2d() methods and learn how they work.

Point-Slope Equation

The graph method that we just wrote is going to come in handy now. We don't need it, but it will make the next two example programs more interesting, so we'll leave it in. Below is the screen capture of the Point-Slope Calculation program.

You may wonder why we care to revisit these nasty equations from high school. These are the uses for math that your teacher never explained. I'm guessing none of your math teachers said, "Learn this stuff, because it will help you write really cool Java programs." That's what they should have said.

Take for instance the point-slope equation of a line. If you're writing a Pong clone, point-slope gives you a simplistic equation for figuring out when the ball hit the wall or paddle. Let's make a graph of user inputted lines.

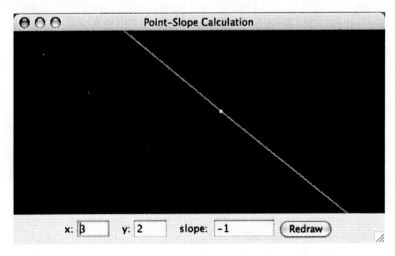

Learning Java Bindings for OpenGL (JOGL)

The main class is:

```java
import java.awt.*;
import java.awt.event.*;
import javax.swing.*;
import net.java.games.jogl.*;

/**
 * This is a basic JOGL app. Feel free to
 * reuse this code or modify it.
 */
public class LineGraphApp extends JFrame implements ActionListener {
    //Notice we've given these two objects a larger scope.
    //Local scope to the constructor was no longer sufficient.
    LineGLEventListener listener = new LineGLEventListener();
    GLCanvas glcanvas;

    JTextField ajtf = new JTextField("3", 3);
    JTextField bjtf = new JTextField("2", 3);
    JTextField mjtf = new JTextField("-1", 6);

    public static void main(String[] args) {
        final LineGraphApp app = new LineGraphApp();

        // show what we've done
        SwingUtilities.invokeLater (
            new Runnable() {
                public void run() {
                    app.setVisible(true);
                }
            }
        );
    }

    public LineGraphApp() {
        //set the JFrame title
        super("Point-Slope Calculation");

        //kill the process when the JFrame is closed
        setDefaultCloseOperation(JFrame.EXIT_ON_CLOSE);
```

```java
        //Setting up our southern JPanel
        JPanel jp = new JPanel();

        //adding the JTextFields and JLabels
        jp.add(new JLabel("x:"));
        jp.add(ajtf);
        jp.add(new JLabel("   y:"));
        jp.add(bjtf);
        jp.add(new JLabel("   slope: "));
        jp.add(mjtf);

        //adding the JButton
        JButton jb = new JButton("Redraw");
        jb.addActionListener(this);
        jp.add(jb);

        getContentPane().add("South", jp);

        //only three JOGL lines of code ... and here they are
        GLCapabilities glcaps = new GLCapabilities();

        glcanvas =
        GLDrawableFactory.getFactory().createGLCanvas(glcaps);

        glcanvas.addGLEventListener(listener);

        //add the GLCanvas just like we would any Component
        getContentPane().add(glcanvas, BorderLayout.CENTER);
        setSize(500, 300);

        //center the JFrame on the screen
        centerWindow(this);
    }

    public void centerWindow(Component frame) {
        Dimension screenSize =
            Toolkit.getDefaultToolkit().getScreenSize();
        Dimension frameSize  = frame.getSize();
```

```
        if (frameSize.width  > screenSize.width )
           frameSize.width   = screenSize.width;
        if (frameSize.height > screenSize.height)
           frameSize.height = screenSize.height;

        frame.setLocation (
            (screenSize.width  - frameSize.width ) >> 1,
            (screenSize.height - frameSize.height) >> 1
        );
    }

    /**
     * Implementation of our ActionListener. This allows the
     * buttons to perform an action. In this case they set
     * the "whatToDraw" String and ask for a repaint of the
     * GLCanvas.
     */
    public void actionPerformed(ActionEvent ae) {
        listener.a = Double.parseDouble( ajtf.getText() );
        listener.b = Double.parseDouble( bjtf.getText() );
        listener.m = Double.parseDouble( mjtf.getText() );
        glcanvas.repaint();
    }
}
```

The GLEventListener follows. Once again we've limited the program to an object with a main() method and a GLEventListener implementation. I hope you're seeing a pattern.

```
import java.awt.*;
import java.awt.event.*;
import net.java.games.jogl.*;

/**
 * For our purposes only two of the GLEventListeners matter. Those would be
```

* init() and display().
 */
public class LineGLEventListener implements GLEventListener {

    //public slope and (a,b) for setting up the line
    public double m = -1;
    public double a = 3;
    public double b = 2;

    //floats used for color selection
    float red;
    float green;
    float blue;

    ///
    //
    // GLEventListener implementation
    //

    /**
     * Take care of initialization here.
     */
    public void init(GLDrawable gld) {
        GL gl = gld.getGL();
        GLU glu = gld.getGLU();

        gl.glClearColor(0.0f, 0.0f, 0.0f, 0.0f);

        gl.glLineWidth(2.0f);
        gl.glPointSize(2.0f);

        gl.glViewport(-250, -150, 250, 150);
        gl.glMatrixMode(GL.GL_PROJECTION);
        gl.glLoadIdentity();
        glu.gluOrtho2D(-250.0, 250.0, -150.0, 150.0);
    }

    /**
     * Take care of drawing here.
     */
    public void display(GLDrawable gld) {
        GL gl = gld.getGL();
```
```

```
drawGraph( gl );

//This is the new code. We find out
//which trig function is selected,
//then we draw a scaled up version of
//the function.

//Let's make the line red
red = 1.0f;
green = 0.2f;
blue = 0.2f;

gl.glColor3f(red, green, blue);

// Point-slope form of a line is:
// y = m(x -a) + b where (a,b) is the
// point.
// Also,
// y - b = m( x - a )
// works.
// m is of course the slope.

//Let's make the line
gl.glBegin(GL.GL_POINTS);

//let's make every grid one point even
//though it is made by 10 x 10 pixels.
double a1 = a * 10;
double b1 = b * 10;

for (int x=-250; x<=250; x++) {
    gl.glVertex2d(x, (m * (x - a1) + b1) );
}

gl.glEnd();


//Let's make the point now

//making the point white
red = 1.0f;
green = 1.0f;
blue = 1.0f;
gl.glColor3f(red, green, blue);
```

```
        gl.glPointSize(4.0f);

        gl.glBegin(GL.GL_POINTS);
        gl.glVertex2d( a1, b1 );
        gl.glEnd();

        //resetting the point to 2 pixels.
        gl.glPointSize(2.0f);
    }


    public void reshape(
                    GLDrawable drawable,
                    int x,
                    int y,
                    int width,
                    int height
                ) {}

    public void displayChanged(
                        GLDrawable drawable,
                        boolean modeChanged,
                        boolean deviceChanged
                    ) {}

    ///////////////////////////////////////////////
//
    // Other methods
    //

    /**
     * In here we draw a Cartesian Coordinate System.
     */
    private void drawGraph(GL gl) {

        gl.glClear(GL.GL_COLOR_BUFFER_BIT);

        //////////////////////
        //drawing the grid
        red = 0.2f;
        green = 0.2f;
        blue = 0.2f;

        gl.glColor3f(red, green, blue);
```

```
//You may notice I'm using GL_LINES here.
//Details of glBegin() will be discussed
  latter.
gl.glBegin(GL.GL_LINES);

//draw the vertical lines
for (int x=-250; x<=250; x+=10) {
    gl.glVertex2d(x, -150);
    gl.glVertex2d(x, 150);
}

//draw the horizontal lines
for (int y=-150; y<=150; y+=10) {
    gl.glVertex2d(-250, y);
    gl.glVertex2d(250, y);
}

gl.glEnd();

//////////////////////////////
// draw the x-axis and y-axis
red = 0.0f;
green = 0.2f;
blue = 0.4f;

gl.glColor3f(red, green, blue);

gl.glBegin(GL.GL_LINES);

//line for y-axis
gl.glVertex2d(0, 140);
gl.glVertex2d(0, -140);

//line for x-axis
gl.glVertex2d(240, 0);
gl.glVertex2d(-240, 0);

gl.glEnd();

////////////////////////
// draw arrow heads
gl.glBegin(GL.GL_TRIANGLES);

gl.glVertex2d( 0, 150);
gl.glVertex2d(-5, 140);
```

```
            gl.glVertex2d( 5, 140);

            gl.glVertex2d( 0, -150);
            gl.glVertex2d(-5, -140);
            gl.glVertex2d( 5, -140);

            gl.glVertex2d(250, 0);
            gl.glVertex2d(240,-5);
            gl.glVertex2d(240, 5);

            gl.glVertex2d(-250, 0);
            gl.glVertex2d(-240,-5);
            gl.glVertex2d(-240, 5);

            gl.glEnd();
      }
}
```

If you were paying attention, you should have noticed that LineGraphApp implements ActionListener. This is to handle input from the button. You can receive events from the GLCanvas too, but we'll be discussing that in detail in another chapter. For now we'll use the more traditional button clicks.

The big change is in the display() method in the LineGLEventListener class. We literally draw the line point by point. We could have drawn it as one line instead of individual points, but where's the fun in that?

glVertex2d() draws the points, but only when it is surrounded by glBegin(GL.GL_POINTS) and a glEnd(). glPointSize(4.0f) is used to make the selected point's size large enough to be seen. We also change the color of the selected point to white. White is red of value 1.0f, green of value 1.0f and blue of value 1.0f.

Again, read the code and type it in and compile it. With any luck you will type it wrong and have to track down a bug or two. You'll never learn so much as when you're tracking down bugs.

Sine, Cosine and Tangents

One thing I could never figure out a real use for in High School, was Trig. Wow. I hated trig. Then I started doing computer graphics and realized something. Sine and cosine and all that have useful applications.

Here we have a program that displays graphed values of sine, cosine and tangent. I've warped them a bit to make them look nicer, but this

Learning Java Bindings for OpenGL (JOGL)

should be a good review of the Math class. You'll also see the graph function again.

```java
import java.awt.*;
import java.awt.event.*;
import javax.swing.*;
import net.java.games.jogl.*;

/**
 * This is a basic JOGL app. Feel free to
 * reuse this code or modify it.
 */
public class TrigGraphApp
    extends JFrame implements ActionListener
{
    //Notice we've given these two objects a larger scope.
    //Local scope to the constructor was no longer sufficient.
    TrigGLEventListener listener = new TrigGLEventListener();
    GLCanvas glcanvas;

    public static void main(String[] args) {
        final TrigGraphApp app = new TrigGraphApp();

        // show what we've done
        SwingUtilities.invokeLater (
            new Runnable() {
                public void run() {
                    app.setVisible(true);
                }
            }
        );
    }

    public TrigGraphApp() {
        //set the JFrame title
        super("Sine, Cosine and Tangent");

        //kill the process when the JFrame is closed
        setDefaultCloseOperation(JFrame.EXIT_ON_CLOSE);

        //Setting up our southern JPanel with
```

JRadioButtons
```
        JPanel jp = new JPanel();
        ButtonGroup bg = new ButtonGroup();

        //setting up the sine JRadioButton
        JRadioButton jrb1 = new JRadioButton("Sine",
true);
        jrb1.setActionCommand("sine");
        jrb1.addActionListener(this);

        //setting up the cosine JRadioButton
        JRadioButton jrb2 = new
JRadioButton("Cosine");
        jrb2.setActionCommand("cosine");
        jrb2.addActionListener(this);

        //setting up the tangent JRadioButton
        JRadioButton jrb3 = new
JRadioButton("Tangent");
        jrb3.setActionCommand("tangent");
        jrb3.addActionListener(this);

        //adding the buttons to the ButtonGroup
        bg.add( jrb1 );
        bg.add( jrb2 );
        bg.add( jrb3 );

        //adding the buttons to the JPanel
        jp.add( jrb1 );
        jp.add( jrb2 );
        jp.add( jrb3 );

        getContentPane().add("South", jp);

        //only three JOGL lines of code ... and here
they are
        GLCapabilities glcaps = new GLCapabilities();

        glcanvas =
        GLDrawableFactory.getFactory().createGLCanvas
(glcaps);

        glcanvas.addGLEventListener(listener);

        //add the GLCanvas just like we would any
```

```
Component
        getContentPane().add(glcanvas,
BorderLayout.CENTER);
        setSize(500, 300);

        //center the JFrame on the screen
        centerWindow(this);
    }

    public void centerWindow(Component frame) {
        Dimension screenSize =
            Toolkit.getDefaultToolkit().getScreenSize(
);
        Dimension frameSize  = frame.getSize();

        if (frameSize.width  > screenSize.width )
           frameSize.width  = screenSize.width;
        if (frameSize.height > screenSize.height)
           frameSize.height = screenSize.height;

        frame.setLocation (
            (screenSize.width  - frameSize.width ) >>
1,
            (screenSize.height - frameSize.height) >>
1
        );
    }

    /**
     * Implementation of our ActionListener. This
allows the
     * buttons to perform an action. In this case
they set
     * the "whatToDraw" String and ask for a repaint
of the
     * GLCanvas.
     */
    public void actionPerformed(ActionEvent ae) {
        listener.whatToDraw = ae.getActionCommand();
        glcanvas.repaint();
    }
}
```

The GLEventListener ....

Gene Davis

```java
import java.awt.*;
import java.awt.event.*;
import net.java.games.jogl.*;

/**
 * For our purposes only two of the GLEventListeners matter. Those would be
 * init() and display().
 */
public class TrigGLEventListener implements GLEventListener {

    public String whatToDraw = "sine";

    /////////////////////////////////////////////////
    //
    // GLEventListener implementation
    //

    /**
     * Take care of initialization here.
     */
    public void init(GLDrawable gld) {
        GL gl = gld.getGL();
        GLU glu = gld.getGLU();

        gl.glClearColor(0.0f, 0.0f, 0.0f, 0.0f);

        gl.glLineWidth(2.0f);
        gl.glPointSize(2.0f);

        gl.glViewport(-250, -150, 250, 150);
        gl.glMatrixMode(GL.GL_PROJECTION);
        gl.glLoadIdentity();
        glu.gluOrtho2D(-250.0, 250.0, -150.0, 150.0);
    }

    /**
     * Take care of drawing here.
     */
    public void display(GLDrawable gld) {
        GL gl = gld.getGL();
        drawGraph( gl );
```

```
//This is the new code. We find out
//which trig function is selected,
//then we draw a scaled up version of
//the function.

float red;
float green;
float blue;

/////////////////////
//drawing the grid
red = 1.0f;
green = 0.2f;
blue = 0.2f;

gl.glColor3f(red, green, blue);

gl.glBegin(GL.GL_POINTS);

if (whatToDraw.equals("sine")) {
    //draw enlarged sine function
    for (int x=-250; x<250;x++)
        gl.glVertex2d(x, (Math.sin(x/
60.0)*100.0));
} else if (whatToDraw.equals("cosine")) {
    //draw enlarged cosine function
    for (int x=-250; x<250;x++)
        gl.glVertex2d(x, (Math.cos(x/
60.0)*100.0));
} else if (whatToDraw.equals("tangent")) {
    //draw enlarged tangent function
    for (int x=-250; x<250;x++)
        gl.glVertex2d(x, (Math.tan(x/
60.0)*30.0));
}

gl.glEnd();

}


public void reshape(
            GLDrawable drawable,
            int x,
```

```
                    int y,
                    int width,
                    int height
                ) {}
    public void displayChanged(
                    GLDrawable drawable,
                    boolean modeChanged,
                    boolean deviceChanged
                ) {}

    //////////////////////////////////////////////
///
    // Other methods
    //

    /**
     * In here we draw a Cartesian Coordinate System.
     */
    private void drawGraph(GL gl) {
        float red;
        float green;
        float blue;

        gl.glClear(GL.GL_COLOR_BUFFER_BIT);

        ///////////////////////
        //drawing the grid
        red = 0.2f;
        green = 0.2f;
        blue = 0.2f;

        gl.glColor3f(red, green, blue);

        //You may notice I'm using GL_LINES here.
        //Details of glBegin() will be discussed
           latter.
        gl.glBegin(GL.GL_LINES);

        //draw the vertical lines
        for (int x=-250; x<=250; x+=10) {
            gl.glVertex2d(x, -150);
            gl.glVertex2d(x, 150);
        }

        //draw the horizontal lines
```

```
for (int y=-150; y<=150; y+=10) {
    gl.glVertex2d(-250, y);
    gl.glVertex2d(250, y);
}

gl.glEnd();

////////////////////////////////
// draw the x-axis and y-axis
red = 0.0f;
green = 0.2f;
blue = 0.4f;

gl.glColor3f(red, green, blue);

gl.glBegin(GL.GL_LINES);

//line for y-axis
gl.glVertex2d(0, 140);
gl.glVertex2d(0, -140);

//line for x-axis
gl.glVertex2d(240, 0);
gl.glVertex2d(-240, 0);

gl.glEnd();

///////////////////////
// draw arrow heads
gl.glBegin(GL.GL_TRIANGLES);

gl.glVertex2d( 0, 150);
gl.glVertex2d(-5, 140);
gl.glVertex2d( 5, 140);

gl.glVertex2d( 0, -150);
gl.glVertex2d(-5, -140);
gl.glVertex2d( 5, -140);

gl.glVertex2d(250, 0);
gl.glVertex2d(240,-5);
gl.glVertex2d(240, 5);

gl.glVertex2d(-250, 0);
gl.glVertex2d(-240,-5);
```

```
            gl.glVertex2d(-240, 5);

            gl.glEnd();

    }

}
```

This is what it should look like:

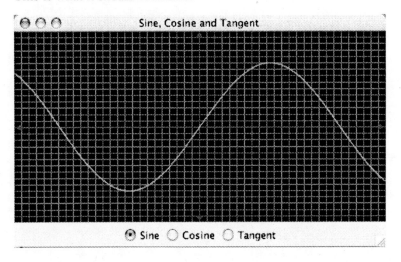

Circles By Hand

Probably the first real intensive piece of graphics I did professionally in Java was a dial or a pie chart. I groaned and wished I had not slept through all those classes. I pulled it off, my boss liked it and I didn't get fired. There's your happily-ever-after for the day.

So here is the secret to making circles. There will be a quiz on this later.

You can find ANY point on the edge of a circle by using the radius measurement and the angle of the point on the circle. If you remember, there are 360 degrees in a circle. The diagram below should serve as a reminder as to where some of the angles are located on the edge of a circle.

Learning Java Bindings for OpenGL (JOGL)

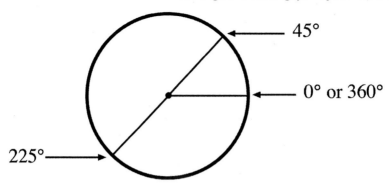

The radius is half the distance across the largest part of the circle. In other words, if you start from the center of the circle and draw a line out to the edge, you've drawn the radius of the circle.

Suppose you decide you want to draw a circle with radius 15 and 360 degrees. (Yes, circles on are defined as having 360 degrees.) We now need to know the x and y coordinates of each of those 360 degrees. Do a for loop that uses this formula:

x = radius * (cosine of angle measured in radians)

and

y = radius * (sine of angle measured in radians)

Hold up! What's a radian? Gotcha.

We're pretty much going to ignore radians, but here is how to define one degree in radians.

```
final double ONE_DEGREE = (Math.PI/180);
```

Likewise, 360 degrees would be,

```
final double THREE_SIXTY = 2 * Math.PI;
```

Drawing a circle using JOGL looks like this.

```
gl.glBegin(GL.GL_POLYGON);
for (double a=0; a<THREE_SIXTY; a+=ONE_DEGREE) {
   x = radius * (Math.cos(a)) + shiftXPosition;
   y = radius * (Math.sin(a)) + shiftYPosition;
   gl.glVertex2d(x, y);
```

}
```
gl.glEnd();
```

Some readers may recognize this as the formula for converting from polar to rectangular coordinates. Good eye. We're just using it for circle creation.

That's about all that is new with the next and final example for this chapter. Again, we'll start by creating the JFrame and placing a GLCanvas in it....

```
import java.awt.*;
import java.awt.event.*;
import javax.swing.*;
import net.java.games.jogl.*;

/**
 * This is a basic JOGL app. Feel free to
 * reuse this code or modify it.
 */
public class FirstCircle extends JFrame {

    public static void main(String[] args) {
        final FirstCircle app = new FirstCircle();

        // show what we've done
        SwingUtilities.invokeLater (
            new Runnable() {
                public void run() {
                    app.setVisible(true);
                }
            }
        );
    }

    public FirstCircle() {
        //set the JFrame title
        super("First Circle Using Sine and Cosine");


        //kill the process when the JFrame is closed
        setDefaultCloseOperation(JFrame.EXIT_ON_CLOSE);

        //only three JOGL lines of code ... and here
```

Learning Java Bindings for OpenGL (JOGL)

```
they are
        GLCapabilities glcaps = new GLCapabilities();

        GLCanvas glcanvas =
        GLDrawableFactory.getFactory().createGLCanvas
(glcaps);

        glcanvas.addGLEventListener(
           new FirstCircleEventListener()
        );

        //add the GLCanvas just like we would any
Component
        getContentPane().add(glcanvas,
BorderLayout.CENTER);
        setSize(500, 300);

        //center the JFrame on the screen
        centerWindow(this);
    }

    public void centerWindow(Component frame) {
        Dimension screenSize =
           Toolkit.getDefaultToolkit().getScreenSize(
);
        Dimension frameSize = frame.getSize();

        if (frameSize.width  > screenSize.width )
           frameSize.width  = screenSize.width;
        if (frameSize.height > screenSize.height)
           frameSize.height = screenSize.height;

        frame.setLocation (
           (screenSize.width  - frameSize.width ) >>
1,
           (screenSize.height - frameSize.height) >>
1
        );
    }
}
```

Now let's display the circle using a GLEventListener --FirstCircleEventListener in this case.

```java
import net.java.games.jogl.*;

/**
 * We make the center of the GLCanvas the origin of our graph
 * and construct a circle around it using sine and cosine
 * methods from the Math class.
 *
 * We have ignored insets in this and other examples, so this
 * circle may be slightly more of an oval depending on the OS.
 */
public class FirstCircleEventListener
  implements GLEventListener {

    final double ONE_DEGREE = (Math.PI/180);
    final double THREE_SIXTY = 2 * Math.PI;

    /**
     * Take care of initialization here.
     */
    public void init(GLDrawable gld) {
        GL gl = gld.getGL();
        GLU glu = gld.getGLU();

        gl.glClearColor(0.0f, 0.0f, 0.0f, 1.0f);

        gl.glViewport(-250, -150, 250, 150);
        gl.glMatrixMode(GL.GL_PROJECTION);
        gl.glLoadIdentity();
        glu.gluOrtho2D(-250.0, 250.0, -150.0, 150.0);
    }

    /**
     * Take care of drawing here.
     */
    public void display(GLDrawable drawable) {
        double x,y;
        double radius = 100;

        float red = 0.5f;
        float green = 0.0f;
        float blue = 0.5f;
```

```
        GL gl = drawable.getGL();

        gl.glClear(GL.GL_COLOR_BUFFER_BIT);

        gl.glColor3f(red, green, blue);

        gl.glBegin(GL.GL_POLYGON);

        // angle is
        // x = radius * (cosine of angle)
        // y = radius * (sine of angle)
        for (double a=0; a<THREE_SIXTY; a+=ONE_DEGREE)
{
            x = radius * (Math.cos(a));
            y = radius * (Math.sin(a));
            gl.glVertex2d(x, y);
        }
        gl.glEnd();
    }

    public void reshape(
                    GLDrawable drawable,
                    int x,
                    int y,
                    int width,
                    int height
                ) {}

    public void displayChanged(
                    GLDrawable drawable,
                    boolean modeChanged,
                    boolean deviceChanged
                ) {}
}
```

It is important to note that we ignore insets of this JFrame, so the circle may appear slightly oval. Other than that, we have just made a perfectly acceptable circle.

Gene Davis

I'm sure wheels are already turning in your mind thinking about some of the things you can do with the JOGL knowledge you've gained so far. But wait! There's more! The whole next chapter is dedicated to looking with more detail at some simple real world examples you can already do with what we've discussed in this chapter. We'll also take a look at animation.

Chapter Three
Animations

Java Thread Review

Hopefully you've used threads before, but I'm going to give you a brief refresher course just in case it is dim in your memory. Threading in pretty much every language comes in two varieties.

One type of threading is to have a timer that acts like an alarm clock. You set it up to go off at some regular interval. In Java, this type of threading can be accomplished with the java.util.Timer class.

Here is a brief example of Timer's use.

```
import java.util.*;

/**
 * This sample shows the creation of a Java
 * Timer object. Timers need a task and a
 * set time to wake up and do that task.
 */
public class TimerExample {

    public static void main (String args[]) {

        //our timer
        Timer alarm = new Timer();

        //our task (really a TimerTask object)
```

```
        TaskExample te = new TaskExample();

        //we schedule the Timer to wake up
        //and do 'te' ever 2000 milliseconds
        //(every two seconds)
        alarm.scheduleAtFixedRate(te, 0, 2000);
    }
}
```

Next we have the actual task to perform.

```
import java.util.*;

/**
 * Timers need a task to perform. Like we need
 * to go to school or work when our alarm
 * clock goes off, the program needs a task
 * to do when it is woken up.
 */
public class TaskExample extends TimerTask {
    /**
     * The task is placed in the run method
     * and performed whenever it is scheduled
     * to do something.
     */
    public void run(){
        System.out.println("Wake up!!!");
    }
}
```

The other type of thread is used to do something right away. It is created in Java by using the java.lang.Thread class.

Here's another example that does the same thing as the Timer we just looked at. This example uses a Thread object and a Runnable implementation to accomplish the task.

```
import java.util.*;

public class ThreadExample {
    /**
     * The main method instantiates a Runnable
     * implementation we called RunnableExample.
     * Then it creates a thread using it.
     * Finally we MUST start() that thread.
```

```
 * run() is called by the JVM after start()
 * is called. Never call run() directly.
 */
public static void main (String args[]) {
    RunnableExample re = new RunnableExample();
    Thread t = new Thread(re);
    t.start();
}
}
```

Next let's look at the sample Runnable implementation.

```
/**
 * We implemented a Runnable here. Some people
 * will say extending the Thread class is more
 * compact. They are correct. That is also bad
 * OOP design, because they are not creating a
 * new Thread type.
 *
 * When it comes down to it. Who cares? It works.
 */
public class RunnableExample implements Runnable {
    public void run() {
        while (true) {
            System.out.println("Wake up!!!");
            try {
                Thread.sleep(2000);
            } catch (Exception e) {}
        }
    }
}
```

Runnable and Repaint

If you have played around with the examples we've been through so far, you may have noticed that you can't animate anything yet. You could set up a Timer or Thread to call repaint() for your GLCanvas frequently, and then update the GLCanvas each repaint. This is actually a common method for animating scenes in Swing and using the AWT.

This kind of effort is wasted as it turns out though. JOGL provides the developer (that would be you) with an Animation class to make animating easy. Let's take a close look at net.java.games.jogl.Animator.

Animation Object

I was about to direct you to look at the Javadoc for the Animator, but realized no one ever actually looks, so let's quote the current version of the doc. It says:

"An Animator can be attached to a GLDrawable to drive its display() method in a loop. For efficiency, it sets up the rendering thread for the drawable to be its own internal thread, so it can not be combined with manual repaints of the surface.

"The Animator currently contains a workaround for a bug in NVidia's drivers (80174). The current semantics are that once an Animator is created with a given GLDrawable as a target, repaints will likely be suspended for that GLDrawable until the Animator is started. This prevents multithreaded access to the context (which can be problematic) when the application's intent is for single-threaded access within the Animator. It is not guaranteed that repaints will be prevented during this time and applications should not rely on this behavior for correctness."

Hmm. Clear as mud. (I've always wanted to say that.) It really isn't that bad. We'll be using the Animator in this chapter to get you used to it and to help you internalize what we've gone over thus far.

The Animator is constructed with a GLDrawable object. Where could we find a handy GLDrawable? As it happens, GLCanvas is a GLDrawable. Now the Animator is associated with the GLDrawable that it was constructed with. That means that if you would like two GLCanvases in your program, but only want one to update itself using an Animator, you can construct the Animator with the GLCanvas you wish to animate and the other GLCanvas will behave normally.

Animator has two methods that we have access to. They are start() and stop(). You've probably already guessed that the start() method starts the animator just as start() would start a Thread object. stop() does just what you would expect it to. It stops the Animator's animation of the GLDrawable. You must call stop() to clean up a running Animator, otherwise your program may not behave properly. However, we won't call stop because we don't care about nice clean up. System.exit(0) will kill the JVM anyway.

Bouncing Off the Walls

Now, let's make a square bounce off the walls of our GLCanvas. We will choose the slope of the line that moves the "ball" around the

Learning Java Bindings for OpenGL (JOGL)

GLCanvas. That will be hard coded in, but feel free to change the slope and point the line is defined by.

For those who don't see the point, think of games like breakout and tennis/pong that require nice bounces. This is a perfect illustration for this kind of game.

We will be using a similar two class setup that we've used for the other illustrations. This time we will be using the Animator class. This will allow the GLCanvas to change on its own, without the user's intervention.

Here's the main() app.

```java
import java.awt.*;
import java.awt.event.*;
import javax.swing.*;
import net.java.games.jogl.*;

public class Bouncing extends JFrame {

    static Animator animator = null;

    public static void main(String[] args) {
        final Bouncing app = new Bouncing();

        // show what we've done
        SwingUtilities.invokeLater (
            new Runnable() {
                public void run() {
                    app.setVisible(true);
                }
            }
        );

        //start the animator
        SwingUtilities.invokeLater (
            new Runnable() {
                public void run() {
                    animator.start();
                }
            }
        );
    }

    public Bouncing() {
        //set the JFrame title
```

```
        super("Bouncing Off the Walls");

        //kill the process when the JFrame is closed
        setDefaultCloseOperation(JFrame.EXIT_ON_
CLOSE);

        //we'll create our GLEventListener
        BouncingDisplay display = new
BouncingDisplay();

        //Now we will create our GLCanvas
        GLCapabilities glcaps = new GLCapabilities();
        GLDrawableFactory gldFactory =
            GLDrawableFactory.getFactory();
        GLCanvas glcanvas = gldFactory.createGLCanvas
(glcaps);
        glcanvas.addGLEventListener(display);

        //create the animator
        animator = new Animator(glcanvas);

        //add the GLCanvas just like we would any
Component
        getContentPane().add(glcanvas,
BorderLayout.CENTER);
        setSize(500, 300);

        //center the JFrame on the screen
        centerWindow(this);
    }

    public void centerWindow(Component frame) {
        Dimension screenSize =
            Toolkit.getDefaultToolkit().getScreenSize(
);
        Dimension frameSize  = frame.getSize();

        if (frameSize.width  > screenSize.width )
            frameSize.width  = screenSize.width;
        if (frameSize.height > screenSize.height)
            frameSize.height = screenSize.height;

        frame.setLocation (
            (screenSize.width  - frameSize.width ) >>
1,
```

```
            (screenSize.height - frameSize.height) >> 1
        );
    }
}
```

The GLEventListener is here.

```
import java.awt.*;
import java.awt.event.*;
import net.java.games.jogl.*;

public class BouncingDisplay implements
GLEventListener {

    float a = 250;//x axis
    float b = 150;//y axis

    //Remember to use floats for calculating slope
    //of the line the ball follows. Ints will be
    //far too imprecise (i.e. (8/9) == 0).
    //
    //Slope will change on each wall impact.
    //It will be multiplied by -1.
    float slope = 7.0f/6.0f;

    float x = a; //holds the new 'x' position of ball
    float y = b; //holds the new 'y' position

    boolean movingRight = true;
    boolean movingUp = true;

    /**
     * Remember that the GLDrawable is actually the
     * GLCanvas that we dealt with earlier.
     */
    public void init(GLDrawable gld) {
        //Remember not to save the
        //GL and GLU objects for
        //use outside of this method.
        //New ones will be provided
        //later.
        GL gl = gld.getGL();
        GLU glu = gld.getGLU();
```

```
        //Let's use a different color than black
        gl.glClearColor(0.725f, 0.722f, 1.0f, 0.0f);

        //Let's make the point 5 pixels wide
        gl.glPointSize(5.0f);

        //For simplicity, let's set the viewport
        //and the coordinate system to display
        //points in the range (0,0) and (500, 300);

        //glViewport's arguments represent
        //left, bottom, width, height
        gl.glViewport(0, 0, 500, 300);
        gl.glMatrixMode(GL.GL_PROJECTION);
        gl.glLoadIdentity();
        //gluOrtho2D's arguments represent
        //left, right, bottom, top
        glu.gluOrtho2D(0, 500, 0, 300);
}
    public void display(GLDrawable gld) {
        // Remember to get a new copy
        // of GL object instead of
        // saving a previous one
        GL gl = gld.getGL();

        //erase GLCanvas using the clear color
        gl.glClear(GL.GL_COLOR_BUFFER_BIT);

        //Choose our color for drawing
        float red = 0.1f;
        float green = 0.5f;
        float blue = 0.1f;
        gl.glColor3f(red, green, blue);

        // Point-slope form of a line is:
        // y = m(x -a) + b where (a,b) is
        // the point.
        //
        // Also,
        // y - b = m( x - a )
        // works.
        // m is of course the slope.

        //(x,y) position of point changes
```

```
//each time this frame is drawn.

y = (slope * (x - a) + b);

//note for our bounce we will
//use the formula:
//slope *= -1

  if (movingRight) {
     if (x < 500) {
        x += .2;
     } else {
        movingRight = false;
        slope *= -1;
        a = x;
        b = y;
     }
}
if (! movingRight) {
    if (x > 0) {
       x -= .2;
    } else {
       movingRight = true;
       slope *= -1;
       a = x;
       b = y;
    }
}

  if (movingUp) {
     if (! (y < 300)) {
        slope *= -1;
        a = x;
        b = y;
        movingUp = false;
     }
}
if (! movingUp) {
    if (! (y > 0)) {
       slope *= -1;
       a = x;
       b = y;
       movingUp = true;
    }
}
```

```
            //only one point (our ball) to draw
            gl.glBegin(GL.GL_POINTS);
               gl.glVertex2d(x, y);
            gl.glEnd();
    }

    //we won't need these two methods
    public void reshape(
                        GLDrawable drawable,
                        int x,
                        int y,
                        int width,
                        int height
                       ) {}

    public void displayChanged(
                               GLDrawable drawable,
                               boolean modeChanged,
                               boolean deviceChanged
                              ) {}

}
```

Hopefully you read the comments as you typed the program in. You did try it, didn't you? For those who may have not paid attention I'll review a few things.

Our ball, in this program travels in a line. We have to know two bits of information in order to calculate that line and place our ball on it. First we need to know a point that is on that line. It doesn't matter where that point is, but we need to know its x and y position. Second we need to know the slope of the line.

What is the slope of the line? In its simplest form slope is defined as rise over run.

$$\text{slope} = \frac{\text{rise}}{\text{run}}$$

If you traveled one unit right in the coordinate system and your line went up by two units you would have a slope of two. That is because two divided by one is two. It also has a positive slope. If the line had gone down by two units, instead of up, then the slope would be -2.

Learning Java Bindings for OpenGL (JOGL)

In our program, we start out with an initial slope and a point to define the line our ball will travel along. When we encounter a wall (GLCanvas edge) we create a new line based off the old line.

To make our change we need a new point for the new line. We use the current point the ball is on. That leaves only a new slope to be created. The new slope is created by multiplying the old slope by -1. This replaces upward slopes with downward or vice versa.

Also note that we created our Animator using a GLCanvas. Then after showing the JFrame containing the GLCanvas, we call the start() method. We would call stop() to stop the animation if we weren't exiting the program.

Rotating a Handmade Dial

I've only had to do full blown dials twice in my career, but that means I needed to understand how it is done. It will benefit you to know how this is done too. I'm going to simplify this for the sake of illustration. We won't accept user input. We'll just have the dial move a predetermined distance.

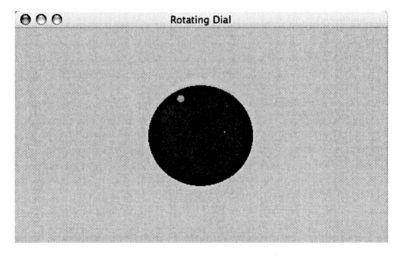

Perception is the name of the game. With movies, computer games and magic tricks, the important thing to do is make your observer think something has happened. That means, you don't have to actually accomplish the task, just make it look like you have.

In a complex dial, you might have to really move every visible point to make it look like it has been turned. On a simple dial such as this one, only one point really moves. You may remember the circle we drew in the last chapter, let's turn it into a dial.

```java
import java.awt.*;
import java.awt.event.*;
import javax.swing.*;
import net.java.games.jogl.*;

public class Dial extends JFrame {

    public static Animator animator = null;

    public static void main(String[] args) {
        final Dial app = new Dial();

        // show what we've done
        SwingUtilities.invokeLater (
            new Runnable() {
                public void run() {
                    app.setVisible(true);
                }
            }
        );

        //start the animator
        SwingUtilities.invokeLater (
            new Runnable() {
                public void run() {
                    animator.start();
                }
            }
        );
    }

    public Dial() {
        super("Rotating Dial");

        //kill the process when the JFrame is closed
        setDefaultCloseOperation(JFrame.EXIT_ON_CLOSE);

        //we'll create our GLEventListener
        DialDisplay display = new DialDisplay();

        //Now we will create our GLCanvas
        GLCapabilities glcaps = new GLCapabilities();
        GLDrawableFactory gldFactory =
            GLDrawableFactory.getFactory();
```

```
        GLCanvas glcanvas = gldFactory.createGLCanvas
(glcaps);
        glcanvas.addGLEventListener(display);

        //create the animator
        animator = new Animator(glcanvas);

        getContentPane().add(glcanvas,
BorderLayout.CENTER);
        setSize(500, 300);

        //center the JFrame on the screen
        centerWindow(this);
    }

    public void centerWindow(Component frame) {
        Dimension screenSize =
            Toolkit.getDefaultToolkit().getScreenSize(
);
        Dimension frameSize = frame.getSize();

        if (frameSize.width  > screenSize.width )
            frameSize.width  = screenSize.width;
        if (frameSize.height > screenSize.height)
            frameSize.height = screenSize.height;

        frame.setLocation (
            (screenSize.width  - frameSize.width ) >>
1,
            (screenSize.height - frameSize.height) >>
1
        );
    }
}
```

The fun part is the GLEventListener.

```
import java.awt.*;
import java.awt.event.*;
import javax.swing.*;
import net.java.games.jogl.*;

public class DialDisplay implements GLEventListener {
    boolean running = true;
```

```java
        final double ONE_DEGREE = (Math.PI/180);
        final double THREE_SIXTY = 2 * Math.PI;
        double angle = 240 * ONE_DEGREE;

        /**
         * Remember that the GLDrawable is actually the
         * GLCanvas that we dealt with earlier.
         */
        public void init(GLDrawable gld) {
            //Remember not to save the
            //GL and GLU objects for
            //use outside of this method.
            //New ones will be provided
            //later.
            GL gl = gld.getGL();
            GLU glu = gld.getGLU();

            //Let's use a different color than black
            gl.glClearColor(0.725f, 0.722f, 1.0f, 0.0f);

            //Let's make the line 5 pixels wide
            //gl.glLineWidth(5.0f);

            //For simplicity, let's set the viewport
            //and the coordinate system to display
            //points in the range (0,0) and (500, 300);

            //glViewport's arguments represent
            //left, bottom, width, height
            gl.glViewport(0, 0, 500, 300);
            gl.glMatrixMode(GL.GL_PROJECTION);
            gl.glLoadIdentity();
            //gluOrtho2D's arguments represent
            //left, right, bottom, top
            glu.gluOrtho2D(0, 500, 0, 300);
        }

        public void display(GLDrawable gld) {
            double x,y;
            double radius = 70;

            int shiftXPosition = 250;
            int shiftYPosition = 150;

            float red = 0.2f;
```

```
            float green = 0.2f;
            float blue = 0.2f;

            GL gl = gld.getGL();

            gl.glClear(GL.GL_COLOR_BUFFER_BIT);

            gl.glColor3f(red, green, blue);

            gl.glBegin(GL.GL_POLYGON);
            // x = radius * (cosine of angle)
            // y = radius * (sine of angle)
            for (double a=0; a<THREE_SIXTY; a+=ONE_DEGREE)
{
                x = radius * (Math.cos(a)) +
shiftXPosition;
                y = radius * (Math.sin(a)) +
shiftYPosition;
                gl.glVertex2d(x, y);
            }
            gl.glEnd();

            red = 1.0f;
            green = 0.2f;
            blue = 0.2f;
            gl.glColor3f(red, green, blue);

            if (angle > (30*ONE_DEGREE)) angle -= ONE_
DEGREE/100;
            else if (running) {
                //stop the animator
                //we're done with it
                SwingUtilities.invokeLater (
                    new Runnable() {
                        public void run() {
                            Dial.animator.stop();
                        }
                    }
                );
                running = false;
            }

            double tmpXShift =
                (radius-12) * (Math.cos(angle)) +
shiftXPosition;
```

```
            double tmpYShift =
                (radius-12) * (Math.sin(angle)) +
shiftYPosition;

            gl.glBegin(GL.GL_POLYGON);
            for (double a=0; a<THREE_SIXTY; a+=ONE_DEGREE)
{
                x = 5 * (Math.cos(a)) + tmpXShift;
                y = 5 * (Math.sin(a)) + tmpYShift;
                gl.glVertex2d(x, y);
            }
            gl.glEnd();
    }

    //we won't need these two methods
    public void reshape(
                    GLDrawable drawable,
                    int x,
                    int y,
                    int width,
                    int height
                ) {}

    public void displayChanged(
                    GLDrawable drawable,
                    boolean modeChanged,
                    boolean deviceChanged
                ) {}

}
```

We've defined a degree again as ONE_DEGREE = (Math.PI/180). Using 360 degrees is much easier than playing directly with 2*Math.PI radians. We have also defined THREE_SIXTY as 2*Math.PI. Any time we want to specify a number of degrees, we multiply the degrees by ONE_DEGREE. Isn't life easy?

The only other tricky thing we did in this program is shift circles around. This kind of shifting is called a translation which is a kind of transformation. In our first circle drawing program, we made the center of the GLCanvas the origin of the coordinate system and the center of the circle. If we hadn't shifted the circle right and up in the coordinate system it would have been drawn in the lower left hand corner with most of it off the GLCanvas.

Transformations are an important tool in JOGL. OpenGL, and hence JOGL, has built in methods for handling complex transformations. For now, we will handle our own transformations. Let's not make learning JOGL too complex for you, but expect to learn matrix manipulation if you're going to become well versed in JOGL or any 3D graphics programming. A good place to start is to pick up a used math textbook at a college bookstore.

Gene Davis

Chapter Four
Using Events
with GLCanvas

Event and Listener Review

A chapter about Events and Listeners belongs in a beginner's JOGL book as much as it would belong in a beginner's AWT or Swing book. So we're going to spend a chapter looking at the nifty possibilities the java.awt.event package provides for us.

If you are a Java guru, then here's what you need to know. GLCanvas and GLJPanel are AWT and Swing classes. Therefore, they behave like other AWT and Swing classes. Don't be afraid to treat them as such. Look over the javadocs for both. GLCanvas extends Canvas and GLJPanel extends JPanel.

Now for the more typical readers, let's start with a brief review of the event-listener model.

Modern Graphic User Interfaces (GUIs) are typically event driven. Ack! Don't worry, no special driver's license is required. When you learned to program you probably learned about the main() method. Everything in the program happened because you told it to happen directly or indirectly from the main() method. This type of program is called a procedure driven program.

GLCanvas and GLJPanel are part of the AWT and Swing. This allows you to provide for event driven applications. Event driven applications have code written that is only called when the user of the program is interested in running it. Some parts of the program may never run, because the user just isn't interested in using that part of the program.

Below are a few typical use cases for an application with a GLCanvas. Notice everything starts with our stick figure, "Program User". He may type on the keyboard. He may move the mouse, or he may click the mouse. What happens next is that an interface we've implemented is notified using an event. The event is used to start the process of executing code we've written. This is why it is often referred to as "Event Driven" code. This model is called the "Event-Listener" model. The hyphen is optional.

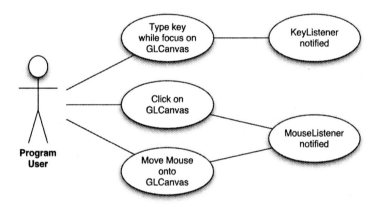

GLCanvas and GLJPanel are both java.awt.Components. This is because of inheritance. The methods that are available to Components are also available to its descendants. Now someone is likely to be thinking of an exception (I am), but this is how things are intended to work.

Usually a GUI Component (as in java.awt.Component) or one of its descendants will be interested in things the user of a program is doing that could affect it. There are many things a user could do; some of the more basic events that a user can cause are dealt with at the Component level .java.util.EventObject and java.util.EventListener are at the heart of the Java event-listener model. When we want a chunk of code to be aware of events that the user has generated, we implement an appropriate Listener. Usually Listeners implement some interface that in turn implements java.util.EventListener. For the Component to know it needs to notify that listener of related events, the Listener needs to be added to the Component using the proper method.

Whenever the user triggers an event that our chunk of code should act on, an event is sent to the properly implemented method. (We implemented that method, ya know.) The event will be a descendant of the java.awt.AWTEvent which in turn will be a child of the java.util.EventObject. You will probably want to glance at the javadocs for these classes.

Let's move on to some concrete examples.

FocusListeners

Let's do our first application with two GLCanvas objects. When we select one with the mouse it will gain focus, and when we select the other, it will gain focus. The focused GLCanvas will lose focus whenever the JFrame or other GLCanvas is selected with the mouse. We will indicate this to the user by changing the color of the GLCanvases.

Here are the two classes first; afterwards, we'll get into the explanations.

```
import java.awt.*;
import java.awt.event.*;
import javax.swing.*;
import net.java.games.jogl.*;

/**
 * This is a basic JOGL app. Feel free to
 * reuse this code or modify it.
 */
public class FocusExample extends JFrame {

  public static void main(String[] args) {
    final FocusExample app = new FocusExample();

    // show what we've done
    SwingUtilities.invokeLater (
      new Runnable() {
        public void run() {
          app.setVisible(true);
        }
      }
    );
  }

  public FocusExample() {
    //set the JFrame title
    super("Focus Example");
```

```java
    //kill the process when the JFrame is closed
    setDefaultCloseOperation(JFrame.EXIT_ON_CLOSE);

    //only three JOGL lines of code ... and here they are
    GLCapabilities glcaps = new GLCapabilities();
    GLCanvas glcanvasWest =
        GLDrawableFactory.getFactory().createGLCanvas(glcaps);
    FocusExampleDisplay fed1 =
        new FocusExampleDisplay("Red", glcanvasWest);
    glcanvasWest.addGLEventListener(fed1);
    glcanvasWest.addFocusListener(fed1);

    GLCanvas glcanvasEast =
        GLDrawableFactory.getFactory().createGLCanvas(glcaps);
    FocusExampleDisplay fed2 =
        new FocusExampleDisplay("Blue", glcanvasEast);
    glcanvasEast.addGLEventListener(fed2);
    glcanvasEast.addFocusListener(fed2);

    getContentPane().setLayout(null);
    setResizable(false);

    //add the GLCanvases just like we would any Components
    getContentPane().add(glcanvasWest, BorderLayout.WEST);
    getContentPane().add(glcanvasEast, BorderLayout.EAST);

    glcanvasEast.setSize(200, 200);
    glcanvasEast.setLocation(275, 50);
    glcanvasWest.setSize(200, 200);
    glcanvasWest.setLocation(25, 50);

    setSize(500, 300);

    //center the JFrame on the screen
    centerWindow(this);
 }

  public void centerWindow(Component frame) {
```

```
   Dimension screenSize =
      Toolkit.getDefaultToolkit().getScreenSize();
   Dimension frameSize  = frame.getSize();

   if (frameSize.width  > screenSize.width )
      frameSize.width  = screenSize.width;
   if (frameSize.height > screenSize.height)
      frameSize.height = screenSize.height;

   frame.setLocation (
     (screenSize.width  - frameSize.width ) >> 1,
     (screenSize.height - frameSize.height) >> 1
   );
  }
}
```

And the actual FocusListener is also our GLEventListener. Look at the bottom of the class for the two methods that implement the Focus Listener.

```
import java.awt.*;
import java.awt.event.*;
import net.java.games.jogl.*;

public class FocusExampleDisplay
   implements GLEventListener, FocusListener
{

  GLCanvas glcanvas = null;

  float redForClearColor   = 0.0f;
  float greenForClearColor = 0.0f;
  float blueForClearColor  = 0.0f;


  public FocusExampleDisplay(String color, GLCanvas
glcanvas) {
     if (color.equals("Red")) redForClearColor = 1.0f;
     else blueForClearColor = 1.0f;
     this.glcanvas = glcanvas;
  }

  /**
   * Take care of initialization here.
```

```java
 */
public void init(GLDrawable drawable) {
  GL gl = drawable.getGL();
  GLU glu = drawable.getGLU();

  gl.glViewport(0, 0, 100, 100);
  gl.glMatrixMode(GL.GL_PROJECTION);
  gl.glLoadIdentity();
  glu.gluOrtho2D(0.0, 100.0, 0.0, 100.0);
}

/**
 * Take care of drawing here.
 */
public void display(GLDrawable drawable) {
  GL gl = drawable.getGL();
  gl.glClearColor(
                  redForClearColor,
                  greenForClearColor,
                  blueForClearColor,
                  1.0f
                 );
  gl.glClear(GL.GL_COLOR_BUFFER_BIT);
}

  public void reshape(
                  GLDrawable drawable,
                  int x,
                  int y,
                  int width,
                  int height
                 ) {}

public void displayChanged(
                  GLDrawable drawable,
                  boolean modeChanged,
                  boolean deviceChanged
                 ) {}

////////////////////////////////
// FocusListener Implementation

public void focusGained(FocusEvent fe) {
  greenForClearColor = 1.0f;
  glcanvas.repaint();
```

```
}

public void focusLost(FocusEvent fe) {
   greenForClearColor = 0.0f;
   glcanvas.repaint();
}
}
```

There are two GLCanvases. Each gets assigned its own instance of FocusExampleDisplay. The FocusExampleDisplay is a GLEventListener and a FocusListener. This is useful, because the GLEventListener is interested in any FocusEvents that are fired.

Our GLEventListener doesn't do anything fancy. It just uses the display() method to clear the canvas to the current clear color. Now the interesting question is how does the clear color get set? The constructor for the FocusExampleDisplay takes a GLCanvas and a String as arguments. The String is used to decide if the default clear color of the GLCanvas should be red or blue. The GLCanvas is used by the Focus listener to cause a repaint to occur.

The focus listener methods change the clear color when they receive a FocusEvent that was fired by some user action. When the focus is gained the green portion of the color is set to 1.0f. When the focus is lost, the green portion of the clear color is set to 0.0f.

Below is a screen capture of the FocusExample application.

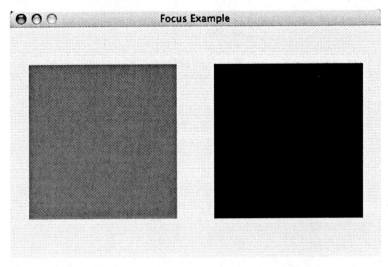

To understand the FocusListener, you will need to know what "focus" is. Focus is short for "keyboard focus". Focus, in windowing toolkits like

Learning Java Bindings for OpenGL (JOGL)

the AWT for Java, is used to describe the Component that you can type in currently. For instance, if you are typing in a TextField or TextArea in a Java program, typically that TextField or TextArea has focus.

So can you type text in a GLCanvas or GLJPanel? Yes and no. Yes, it can be done but displaying the text is unpleasant. I'd recommend using a Component that actually was meant for displaying text for any text intensive applications. However if your goal is to just know what has been typed, then you're in luck. Capturing KeyEvents is reasonably easy, and we discuss KeyEvents later in this chapter.

Two methods must be implemented to make a class a FocusListener: First, the focusGained() method; and second, the focusLost() method.

Focus is gained when a mouse is clicked in the GLCanvas. After the focus is gained, anything typed in from the keyboard is handled by the GLCanvas or other focused Component. Some operating systems allow keyboard navigation. This allows the user to press a key such as the tab key and change what has focus. So, keyboard navigation is another way for Components, like

GLCanvases to gain focus.

Focus is lost when the user clicks somewhere on the screen that doesn't currently have focus, or uses the keyboard to navigate away from the currently focused Component. Once focus is lost from the GLCanvas, the GLCanvas will not know that the user has typed anything on the keyboard.

Both the focusGained() and the focusLost() methods take FocusEvent as an argument. If you have only one FocusListener listening to several Components, then FocusEvent can be handy for determining which Component has changed focus state. The method getComponent() returns the Component that sent the event.

KeyListeners

Now we are going to look at a KeyListener with a GLCanvas. Let's start with the code.

The main class:

```
import java.awt.*;
import java.awt.event.*;
import javax.swing.*;
import net.java.games.jogl.*;

/**
 * This is a basic JOGL app. Feel free to
```

```
 * reuse this code or modify it.
 */
public class KeyExample extends JFrame {

  static GLCanvas glcanvas = null;

  public static void main(String[] args) {
    final KeyExample app = new KeyExample();

    // show what we've done
    SwingUtilities.invokeLater (
      new Runnable() {
        public void run() {
          app.setVisible(true);
          glcanvas.requestFocusInWindow();
        }
      }
    );
  }

  public KeyExample() {
    //set the JFrame title
    super("KeyListener Example");

    //kill the process when the JFrame is closed
    setDefaultCloseOperation(JFrame.EXIT_ON_CLOSE);

    //create our KeyDisplay which serves two purposes
    // 1) it is our GLEventListener, and
    // 2) it is our KeyListener
    KeyDisplay kd = new KeyDisplay();

    //only three JOGL lines of code ... and here they are
    GLCapabilities glcaps = new GLCapabilities();
    glcanvas =
        GLDrawableFactory.getFactory().createGLCanvas(glcaps);
    glcanvas.addGLEventListener(kd);
    glcanvas.addKeyListener(kd);

    //we'll want this for our repaint requests
    kd.setGLCanvas(glcanvas);

    //add the GLCanvas just like we would any
```

```
Component
    getContentPane().add(glcanvas,
BorderLayout.CENTER);
    setSize(500, 300);

    //center the JFrame on the screen
    centerWindow(this);
  }

  public void centerWindow(Component frame) {
    Dimension screenSize =
       Toolkit.getDefaultToolkit().getScreenSize();
    Dimension frameSize  = frame.getSize();

    if (frameSize.width > screenSize.width )
       frameSize.width  = screenSize.width;
    if (frameSize.height > screenSize.height)
       frameSize.height = screenSize.height;

    frame.setLocation (
       (screenSize.width  - frameSize.width ) >> 1,
       (screenSize.height - frameSize.height) >> 1
    );
  }
}
```

Now for the GLEventListener and KeyListener. They are both implemented by one class as the FocusListener and GLEventListener were in the last section.

```
import java.awt.*;
import java.awt.event.*;
import net.java.games.jogl.*;

/**
 * For our purposes only two of the
 * GLEventListeners matter. Those would
 * be init() and display().
 */
public class KeyDisplay implements GLEventListener,
KeyListener {

  int xPosition = 25;
  int yPosition = 25;
```

```
float red = 0.0f;
float green = 0.0f;
float blue = 1.0f;

GLCanvas glc;

public void setGLCanvas(GLCanvas glc) {
  this.glc = glc;
}

/**
 * Take care of initialization here.
 */
public void init(GLDrawable drawable) {
  GL gl = drawable.getGL();
  GLU glu = drawable.getGLU();

  red = 0.0f;
  green = 0.0f;
  blue = 1.0f;

  gl.glClearColor(red, green, blue, 0.0f);

  gl.glViewport(0, 0, 50, 50);
  gl.glMatrixMode(GL.GL_PROJECTION);
  gl.glLoadIdentity();
  glu.gluOrtho2D(0.0, 50.0, 0.0, 50.0);
}

/**
 * Take care of drawing here.
 */
public void display(GLDrawable drawable) {
  GL gl = drawable.getGL();

  gl.glClear(GL.GL_COLOR_BUFFER_BIT);

  //Remember point size refers
  //to pixels, not the coordinate
  //system we've set up in the
  //GLCanvas
  gl.glPointSize(6.0f);

  red = 0.0f;
```

```
   green = 1.0f;
   blue = 0.0f;

   gl.glColor3f(red, green, blue);

   gl.glBegin(GL.GL_POINTS);
      gl.glVertex2i(xPosition, yPosition);
   gl.glEnd();
}

/**
 * Called when the GLDrawable (GLCanvas
 * or GLJPanel) has changed in size. We
 * won't need this, but you may eventually
 * need it -- just not yet.
 */
public void reshape(
                    GLDrawable drawable,
                    int x,
                    int y,
                    int width,
                    int height
                   ) {}

/**
 * If the display depth is changed while the
 * program is running this method is called.
 * Nowadays this doesn't happen much, unless
 * a programmer has his program do it.
 */
public void displayChanged(
                           GLDrawable drawable,
                           boolean modeChanged,
                           boolean deviceChanged
                          ) {}

/////////////////////////////////////
// KeyListener implementation below

public void keyPressed(KeyEvent e) {}

public void keyReleased(KeyEvent e) {}

public void keyTyped(KeyEvent e) {
   if (e.getKeyChar() == KeyEvent.VK_8)
```

```
            yPosition += 1;
         else if (e.getKeyChar() == KeyEvent.VK_2)
            yPosition -= 1;
         else if (e.getKeyChar() == KeyEvent.VK_4)
            xPosition -= 1;
         else if (e.getKeyChar() == KeyEvent.VK_6)
            xPosition += 1;

         glc.repaint();
   }
}
```

You'll notice that this example like many of the other examples in the book is made of two classes. Obviously a useful application is likely to be made of many classes, but simplicity serves us well.

You'll also notice that the main class sets a request for focus for the GLClass. This needs to be done after the request for setVisible(true). You will fail if you try to get focus for a Component that is not visible.

The KeyExample constructor adds the GLEventListener to the GLCanvas as normal, but the KeyDisplay class is also our KeyListener, so we need to add it to the GLCanvas a second time. The second time we add the KeyDisplay to the GLCanvas we use the addKeyListener() method.

KeyListener is an interface to implement if you are interested in knowing about keystrokes. You will only receive notice of keystrokes performed when the Component you subscribe to is in focus. This is why we requested the focus for the GLCanvas object we created. We discussed focus in the last section.

Three methods need to be implemented when creating a KeyListener. They are keyPressed(), keyReleased() and keyTyped(). Each of these takes a KeyEvent object as an argument. KeyEvent gives you more details about the event that was fired.

Once you create a KeyListener implementation and construct an instance, you need to register it with your GLCanvas or GLJPanel. You do this with the addKeyListener() method. Now every time a key is pressed or released you will get KeyEvents fired to your interested implementation of KeyListener. keyTyped() is the method that is most commonly used. If you want to know when a specific letter or number is typed this is likely the method that you're interested in. It is the one we use in our example code too.

A quick look over the KeyEvent class will reveal a host of static attributes. Many of these start with "VK". VK stands for "virtual key". Some examples are VK_5, VK_F12 and VK_ENTER. The virtual keys are useful for comparing the value returned by getKeyCode() and a specific key stroke sought. Virtual keys can often be used with getKeyChar(), but this is more limited in its use.

You may also find the getKeyChar() method useful for returning the specific char primitive associated with the key typed. The getKeyChar() method should not be used with KeyEvents returned from keyPressed() and keyReleased(). It is meant to be used with the keyTyped() method only.

The letters represented by the virtual keys are all uppercase, so will not work for comparing lower case keystrokes returned by getKeyChar() in the keyTyped() method.

Virtual keys show their real power with a getKeyCode() method when used in keyPressed() and keyReleased() methods. You can check for equality in a chunk of code like this:

```
public void keyReleased(KeyEvent e) {
  if ( e.getKeyCode() == KeyEvent.VK_F )
    System.out.println(
    KeyEvent.getKeyText(e.getKeyCode())
  );
}
```

As shown above, you may retrieve the String equivalent of the key code using the getKeyText() method.

It is important to realize the getKeyCode() method is for use in the keyReleased() and keyPressed() methods. It will not be of use in the keyTyped() method. So, use the getKeyChar() method with the KeyEvent in the keyTyped() method, and the getKeyCode() method in the keyReleased() and keyPressed() methods.

MouseListeners

Mouse listeners implement five methods. They are mouseClicked(), mouseEntered(), mouseExited(), mousePressed() and mouseReleased(). mouseClicked() is the most used method. You may find you never need the other four methods.

Like its little brother, MouseMotionListener, the MouseListener receives MouseEvents. MouseEvents contain the x and y coordinate that the mouse caused an event to be fired at. getX() and getY() return int values representing the position of the mouse in the component in the component's coordinate system.

Notice that the Java component coordinate system does not necessarily match the JOGL (OpenGL) coordinate system used for drawing in the GLCanvas. We convert from one to the other in the mouseClicked() method.

Also as all children of ComponentEvents, our event has the originating component as an attribute. We retrieve it with the call e.getComponent(). Someone surely at this point has realized that I could have skipped passing a copy of the GLCanvas to the display method, because the event in the listener method has it already.

Good point, but I wanted the code to be easier to understand. Here is how the repaint could have been done (assuming e is the event) :

```
e.getComponent().repaint();
```

That simplifies the code a little too.

Below is the MouseListener example in the two classes that should be familiar to you by now.

```
import java.awt.*;
import java.awt.event.*;
import javax.swing.*;
import net.java.games.jogl.*;

/**
 * This is a basic JOGL app. Feel free to
 * reuse this code or modify it.
 */
public class MouseExample extends JFrame {

  static GLCanvas glcanvas = null;

  public static void main(String[] args) {
    final MouseExample app = new MouseExample();

    // show what we've done
    SwingUtilities.invokeLater (
      new Runnable() {
        public void run() {
          app.setVisible(true);
        }
      }
    );
```

```java
  }

  public MouseExample() {
    //set the JFrame title
    super("KeyListener Example");

    //kill the process when the JFrame is closed
    setDefaultCloseOperation(JFrame.EXIT_ON_CLOSE);

    //create our KeyDisplay which serves two purposes
    // 1) it is our GLEventListener, and
    // 2) it is our KeyListener
    MouseDisplay md = new MouseDisplay();

    //only three JOGL lines of code ... and here they are
    GLCapabilities glcaps = new GLCapabilities();
    glcanvas =
        GLDrawableFactory.getFactory().createGLCanvas(glcaps);
    glcanvas.addGLEventListener(md);
    glcanvas.addMouseListener(md);

    //we'll want this for our repaint requests
    md.setGLCanvas(glcanvas);

    //add the GLCanvas just like we would any Component
    getContentPane().add(glcanvas, BorderLayout.CENTER);
    setSize(500, 300);

    //center the JFrame on the screen
    centerWindow(this);
  }

  public void centerWindow(Component frame) {
    Dimension screenSize =
        Toolkit.getDefaultToolkit().getScreenSize();
    Dimension frameSize  = frame.getSize();

    if (frameSize.width  > screenSize.width )
       frameSize.width  = screenSize.width;
    if (frameSize.height > screenSize.height)
       frameSize.height = screenSize.height;
```

```
      frame.setLocation (
        (screenSize.width  - frameSize.width ) >> 1,
        (screenSize.height - frameSize.height) >> 1
      );
  }
}

import java.awt.*;
import java.awt.event.*;
import net.java.games.jogl.*;

/**
 * For our purposes only two of the
 * GLEventListeners matter. Those would
 * be init() and display().
 */
public class MouseDisplay
    implements GLEventListener, MouseListener
{

  int xPosition = 50;
  int yPosition = 50;

  float red = 0.0f;
  float green = 0.5f;
  float blue = 0.5f;

  GLCanvas glc;

  public void setGLCanvas(GLCanvas glc) {
    this.glc = glc;
  }

  /**
   * Take care of initialization here.
   */
  public void init(GLDrawable drawable) {
    GL gl = drawable.getGL();
    GLU glu = drawable.getGLU();

    red = 0.0f;
    green = 0.7f;
    blue = 0.3f;
```

```
    gl.glClearColor(red, green, blue, 0.0f);

    gl.glViewport(0, 0, 100, 100);
    gl.glMatrixMode(GL.GL_PROJECTION);
    gl.glLoadIdentity();
    glu.gluOrtho2D(0.0, 100.0, 0.0, 100.0);
}

/**
 * Take care of drawing here.
 */
public void display(GLDrawable drawable) {
    GL gl = drawable.getGL();

    gl.glClear(GL.GL_COLOR_BUFFER_BIT);

    //Remember point size refers
    //to pixels, not the coordinate
    //system we've set up in the
    //GLCanvas
    gl.glPointSize(6.0f);

    red = 0.0f;
    green = 0.0f;
    blue = 0.0f;

    gl.glColor3f(red, green, blue);

    gl.glBegin(GL.GL_POINTS);
        gl.glVertex2i(xPosition, yPosition);
    gl.glEnd();
}

/**
 * Called when the GLDrawable (GLCanvas
 * or GLJPanel) has changed in size. We
 * won't need this, but you may eventually
 * need it -- just not yet.
 */
public void reshape(
                    GLDrawable drawable,
                    int x,
                    int y,
                    int width,
                    int height
```

) {}

```java
/**
 * If the display depth is changed while the
 * program is running this method is called.
 * Nowadays this doesn't happen much, unless
 * a programmer has his program do it.
 */
public void displayChanged(
                           GLDrawable drawable,
                           boolean modeChanged,
                           boolean deviceChanged
                          ) {}

/////////////////////////////////////////////
// MouseListener implementation below

public void mouseClicked(MouseEvent e) {
   double x = e.getX();
   double y = e.getY();

   Component c = e.getComponent();

   double width = c.getWidth();
   double height = c.getHeight();

   //get percent of GLCanvas instead of
   //points and then converting it to our
   //'100' based coordinate system.
   xPosition = (int) ((x / width) * 100);
   yPosition = ((int) ((y / height) * 100));

   //reversing direction of y axis
   yPosition = 100 - yPosition;

   glc.repaint();
}

public void mouseEntered(MouseEvent e) {}
public void mouseExited(MouseEvent e) {}
public void mousePressed(MouseEvent e) {}
public void mouseReleased(MouseEvent e) {}
}
```

You'll notice that using the MouseMotionListener uses pretty much the same code in a new listener implementation. Nice and convenient really.

MouseMotionListeners

Now, we are going to explore a use for mouse motion listeners. Mouse motion listeners implement two methods. One is meant to capture dragging and the other is meant to capture any movement of the mouse.

At this point everything I'm doing should be old hat. I've passed a copy of the GLCanvas to the display object for use in calling repaint. I've added the display object to the GLCanvas as a GLEventListener and as a MouseMotionListener.

Once again, I've converted the position of the mouse pointer as a percentage across the GLCanvas. Then I've converted it to the GLCavas's OpenGL coordinate system. This was set up in the init() method.

Here are the two classes.

```
import java.awt.*;
import java.awt.event.*;
import javax.swing.*;
import net.java.games.jogl.*;

/**
 * This is a basic JOGL app. Feel free to
 * reuse this code or modify it.
 */
public class MouseMotionExample extends JFrame {

  static GLCanvas glcanvas = null;

  public static void main(String[] args) {
    final MouseMotionExample app = new
MouseMotionExample();

    // show what we've done
    SwingUtilities.invokeLater (
      new Runnable() {
        public void run() {
          app.setVisible(true);
          glcanvas.requestFocusInWindow();
        }
      }
```

```
    );
}

public MouseMotionExample() {
  //set the JFrame title
  super("KeyListener Example");

  //kill the process when the JFrame is closed
  setDefaultCloseOperation(JFrame.EXIT_ON_CLOSE);

  //create our KeyDisplay which serves two purposes
  // 1) it is our GLEventListener, and
  // 2) it is our KeyListener
  MouseMotionDisplay mmd = new MouseMotionDisplay();

  //only three JOGL lines of code ... and here they are
  GLCapabilities glcaps = new GLCapabilities();
  glcanvas =
      GLDrawableFactory.getFactory().createGLCanvas(glcaps);
  glcanvas.addGLEventListener(mmd);
  glcanvas.addMouseMotionListener(mmd);

  //we'll want this for our repaint requests
  mmd.setGLCanvas(glcanvas);

  //add the GLCanvas just like we would any Component
  getContentPane().add(glcanvas, BorderLayout.CENTER);
  setSize(500, 300);

  //center the JFrame on the screen
  centerWindow(this);
}

public void centerWindow(Component frame) {
  Dimension screenSize =
      Toolkit.getDefaultToolkit().getScreenSize();
  Dimension frameSize  = frame.getSize();

  if (frameSize.width > screenSize.width )
      frameSize.width  = screenSize.width;
  if (frameSize.height > screenSize.height)
```

```
      frameSize.height = screenSize.height;

    frame.setLocation (
      (screenSize.width  - frameSize.width ) >> 1,
      (screenSize.height - frameSize.height) >> 1
    );
  }
}

import java.awt.*;
import java.awt.event.*;
import net.java.games.jogl.*;

/**
 * For our purposes only two of the
 * GLEventListeners matter. Those would
 * be init() and display().
 */
public class MouseMotionDisplay
   implements GLEventListener, MouseMotionListener
{

  int xPosition = 50;
  int yPosition = 50;

  float red   = 0.0f;
  float green = 0.5f;
  float blue  = 0.5f;

  GLCanvas glc;

  public void setGLCanvas(GLCanvas glc) {
    this.glc = glc;
  }

  /**
   * Take care of initialization here.
   */
  public void init(GLDrawable drawable) {
    GL  gl  = drawable.getGL();
    GLU glu = drawable.getGLU();

    red   = 0.5f;
    green = 0.0f;
    blue  = 1.0f;
```

```java
    gl.glClearColor(red, green, blue, 0.0f);

    gl.glViewport(0, 0, 100, 100);
    gl.glMatrixMode(GL.GL_PROJECTION);
    gl.glLoadIdentity();
    glu.gluOrtho2D(0.0, 100.0, 0.0, 100.0);
}

/**
 * Take care of drawing here.
 */
public void display(GLDrawable drawable) {
    GL gl = drawable.getGL();

    gl.glClear(GL.GL_COLOR_BUFFER_BIT);

    //Remember point size refers
    //to pixels, not the coordinate
    //system we've set up in the
    //GLCanvas
    gl.glPointSize(6.0f);

    red = 0.0f;
    green = 0.0f;
    blue = 0.0f;

    gl.glColor3f(red, green, blue);

    gl.glBegin(GL.GL_POINTS);
        gl.glVertex2i(xPosition, yPosition);
    gl.glEnd();
}

/**
 * Called when the GLDrawable (GLCanvas
 * or GLJPanel) has changed in size. We
 * won't need this, but you may eventually
 * need it -- just not yet.
 */
public void reshape(
                    GLDrawable drawable,
                    int x,
                    int y,
                    int width,
```

```
                    int height
                ) {}

/**
 * If the display depth is changed while the
 * program is running this method is called.
 * Nowadays this doesn't happen much, unless
 * a programmer has his program do it.
 */
public void displayChanged(
                    GLDrawable drawable,
                    boolean modeChanged,
                    boolean deviceChanged
                ) {}

/////////////////////////////////////////////
// MouseMotionListener implementation below

public void mouseDragged(MouseEvent e) {}

public void mouseMoved(MouseEvent e) {
   double x = e.getX();
   double y = e.getY();

   Component c = e.getComponent();

   double width = c.getWidth();
   double height = c.getHeight();

   //get percent of GLCanvas instead of
   //points and then converting it to our
   //'100' based coordinate system.
   xPosition = (int) ((x / width) * 100);
   yPosition = ((int) ((y / height) * 100));

   //reversing direction of y axis
   yPosition = 100 - yPosition;

   glc.repaint();
  }
}
```

Taking the knowledge you've gained in these first four chapters, you can now make some kick-butt applications, ... BUT WAIT, ... there's more.

Gene Davis

Next we'll take our discussion 3D. This is where things get real fun boys and girls. Hang on tight and keep your arms and legs inside the ride!

Chapter Five
Going 3D

What is the Matrix?

    You will want to become very familiar with matrices and matrix math if you want to become a guru of 3D programming. This holds true for JOGL and OpenGL. I'll give you a very brief introduction, but plan on picking up a used math text or two and learning as much as you can. The internet has quite a few tutorials on matrix theory and math, but I'd recommend a book. I haven't seen any tutorials for matrices that impressed me on the net -- not yet anyway.

    All images you see rendered from JOGL are really a bunch of matrices manipulating each other in a fashion set forward by OpenGL. If you've seen any of the Matrix movies or the screen savers based off them, you've seen a matrix. Here is a typical matrix of xyz coordinates:

$$\begin{bmatrix} 12 & 5 & -10 \\ 3 & 14 & 0 \\ 34 & 13 & 4 \\ -5 & -1 & -13 \end{bmatrix}$$

Imagine that the above matrix represents four points in a three dimensional coordinate system. The points would be (12, 5, -10), (3, 14, 0), (34, 13, 4) and (-5, -1, -13).

In Java terms, think of a matrix as an array of arrays. For instance, double d[][] would be used to make a matrix of double primitives. Also a single dimensioned array may be used if you wish to simulate the format of double dimension arrays as they behave in C.

It is important to realize the difference in C and Java when it comes to multi-dimensional arrays. Technically Java doesn't have them. In C, a two dimensional array is all saved together in memory, and can be accessed using pointer math as though it is one array. This is very similar to how Lists are accessed in Java. You will find this to be an important difference to keep in mind as you learn more advanced JOGL and look at OpenGL examples to learn more.

For simpler JOGL work, you won't need to manipulate matrices directly. You'll use methods to do the dirty work for you.

You'll hear the term, 'identity' used a lot with matrices. If you multiply any number by 1 you will end up with itself. Matrices also have an equivalent operation. Instead of multiplying the matrix by 1 to get itself you multiply it by its 'identity'. Identities differ from matrix to matrix, though they are easy to recognize. They all look something like this:

$$\begin{bmatrix} 1 & 0 & 0 \\ 0 & 1 & 0 \\ 0 & 0 & 1 \end{bmatrix}$$

The two main matrices that you will manipulate in JOGL are the GL_PROJECTION matrix and the GL_MODELVIEW matrix. There are others, but you won't need them until you are doing advanced JOGL work.

Setting Matrix Modes

All JOGL based programs have statements similar to:

```
gl.glMatrixMode(GL.GL_PROJECTION);
```

or

```
gl.glMatrixMode(GL.GL_MODELVIEW);
```

The GL class's glMatrixMode() method is used for setting the current matrix mode to manipulate. There are two matrix modes that are commonly used and a host of ones that are only used in special situations, such as for vendor specific features. As mentioned in the last section, the common matrix modes are GL_PROJECTION and GL_MODELVIEW. These are both attributes found in the GL class.

You've probably noticed already that the GL class is monstrous in its size. The javadoc for that class can take some noticeable time to display on even a fast computer. That is one of the small drawbacks of translating C libraries into Java classes.

Think of GL_PROJECTION as your camera mode and GL_MODELVIEW as your everything else mode. There are some methods that should be called only in GL_PROJECTION matrix mode. These methods are:

gluPerspective()
glFrustum() //commonly misspelled glFrustrum
glOrtho()
gluOrtho2D
glLoadIdentity

You could use glLoadMatrix() too, but you won't want to do that until you are very experienced. Rumor has it that if you call that method, you wake up in a futuristic world where our reality is only a computer induced dream, and the food all tastes like rehydrated mashed potatoes with no salt or butter. Mind you, that is just a rumor.

So if you're using one of the methods I've listed, chances are that you should have set the matrix mode to GL_PROJECTION otherwise you should be using the GL_MODELVIEW matrix mode. Luckily GL_MODELVIEW is the default matrix mode.

Typically changing the matrix mode is slow and should be done sparingly.

When you call glMatrixMode() the next call that you will likely want to do is glLoadIdentity(). We mentioned identities earlier in the chapter.

glLoadIdentity() initialized the the matrix we've just loaded to its identity matrix.

I Spy with My Little Eye

How do you see in a 3D world? Everything is projected onto the computer screen. Perspective projection and orthographic projection to the screen can be obtained in more than one way. We'll stick to using gluOrtho2D() for orthographic projection and will introduce you to the use of gluPerspective() for perspective projection on to the screen.

Orthographic projection results in pictures where all the objects are presented as though they don't appear smaller as they get farther away. Orthographic projection lends itself to making pictures look flat. This is useful in many applications, such as CAD programs or drawing sheet music. In an orthographic projection all objects of equal size look like they are the same size, even if one is close to the camera view and the other is far from the camera view. If the old Sesame Street skit with Grover saying "near" and "far" were done in an orthographic view, all the children would be very confused.

Looking at the image below, you may assume that you are looking at three squares. You are in fact looking at three boxes, but cannot tell because you are seeing them in an orthographic view.

Perspective projection results in objects that are farther away from the camera appearing smaller. Also all parallel lines receding from you appear to converge in the distance. Hence the illusion of depth is generated. Most people can appreciate depth added to a screen in this fashion. By viewing the three boxes seen earlier in a perspective view, they suddenly *look* like boxes.

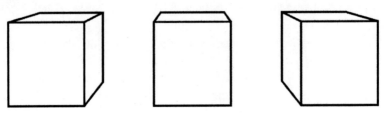

gluLookAt() and gluPerspective()

If a tree falls in a forest and no one is there to see it, was it green and brown? No. It didn't even reflect any light, if it was modeled in JOGL.

Look around at the "real" world. How far can you see? On a clear night you can see stars and galaxies whose light did not reach earth for millions of years. In nature, a starry night being the perfect example, the views you see took nature more than your lifetime to create. All this light finally reaches your eyes at just the right moment, so that you and I can blissfully be ignorant that nature's rendering of a simple nightscape took millions of years.

Why does it take so long for nature to create something as simple as a starry night? It is simply the distances involved. If you tried to duplicate nature's method of rendering a starry night you would quickly realize that there doesn't exist enough computing power on the planet and likely never will to render scenes exactly like nature does it.

Also if I design a scene for my user to view, but throw up a status bar saying it will take thirty years to make, I might not have a customer any more.

The way we get around size and time restrictions is to only create what we want the user to see, and put that in a bounded area called a frustum. What is a frustum? Imagine the view from a camera. If the view were a pyramid on its side with the former top point touching the camera lens, a frustum would be the pyramid with the part near the camera chopped off. We define a frustum so that the computer does not have to worry about infinitely distant objects, or objects that can't be seen. The computer has a fixed set of points to render in the view.

Gene Davis

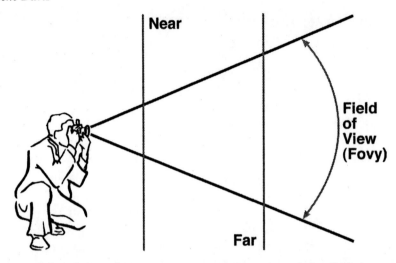

First we decide what the users field of view is. We call this "fovy." Typically for a good perspective from a monitor, 60 degrees works. Second we decide on the aspect. Aspect is the width of the view divided by the height of the view, unless we're looking for a warping effect. Lastly, we care about the distances from the camera that we wish to see. Anything nearer than what we specify or farther than we specify is ignored as though it isn't there. We could see through a wall if the wall is closer to the camera than we've specified in our near argument.

gluPerspective() is positioned relative to the camera position we define with gluLookAt(). That's not too clear, but I'll explain more once we go over the gluLookAt() method. gluPerspective() is defined:

```
gluPerspective(
   double fovy,
   double aspect,
   double near,
   double far
)
```

Near and far need to be positive numbers, never negative. Also, far must be greater than near. This method is purely for setting up the camera. None of the arguments that we pass in to gluPerspective() are actual coordinates. Near and far are distances from the camera lens.

Having said all that, where is the camera? We specify the camera position with the gluLookAt() method. When we set up the camera, we set it up using three points mapped to a three dimensional coordinate system. You define a point for the camera position, a point it is looking at (usually in the middle of your scene) and a point relative to the origin that is considered "up." These points are commonly called "eye", "at" and "up."

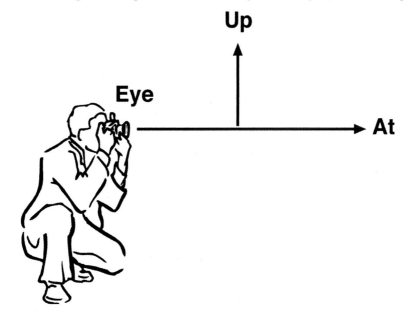

A warning is in order. If you try to define up relative to the camera position instead of the origin, then you will have the camera rotated wrong. Think of the origin as the center of the universe, and up is where the point defined as up is found relative to the origin.

gluLookAt() has the following parameters.

```
gluLookAt(
   double eyeX, double eyeY, double eyeZ,
   double atX, double atY, double atZ,
   double upX, double upY, double upZ
)
```

Now let's get back to the relation of gluLookAt() to gluPerspective().

It is very important to note that the gluPerspective() method does not take any coordinates as arguments. The natural tendency of newbies is to think of near and far as though they are x, y or z values. They are not. The near and far values passed to gluPerspective() are distances from the 'eye' point specified in the gluLookAt() method toward the 'at' point specified in the gluLookAt() method.

Reread the last paragraph until you understand it. This is very important to understand. gluLookAt() specifies where you and your camera are and what you are looking at. gluPerspective() specifies how near and far you can see in the direction gluLookAt() specified.

Three Quads in Perspective

It's time to put all this together in an example. This example shows three quads. In English that would be three squares.

The three squares appear to be different sizes, because they are found at three different distances from the camera we set up with gluLookAt().

```
import java.awt.*;
import java.awt.event.*;
import javax.swing.*;
import net.java.games.jogl.*;

/**
 * This is a basic JOGL app. Feel free to
 * reuse this code or modify it.
 */
public class Planes extends JFrame {

  public static void main(String[] args) {
    final Planes app = new Planes();

    //show what we've done
    SwingUtilities.invokeLater (
      new Runnable() {
        public void run() {
          app.setVisible(true);
        }
      }
    );
  }

  public Planes() {
    //set the JFrame title
    super("Three Planes Application");

    //kill the process when the JFrame is closed
    setDefaultCloseOperation(JFrame.EXIT_ON_CLOSE);

    //only three JOGL lines of code ... and here they are
    GLCapabilities glcaps = new GLCapabilities();
```

```java
      GLCanvas glcanvas =
          GLDrawableFactory.getFactory().createGLCanvas(
glcaps);
      glcanvas.addGLEventListener(new PlanesView());

      //add the GLCanvas just like we would any
Component
      getContentPane().add(glcanvas,
BorderLayout.CENTER);
      setSize(500, 300);

      //center the JFrame on the screen
      centerWindow(this);
   }

   public void centerWindow(Component frame) {
      Dimension screenSize =
          Toolkit.getDefaultToolkit().getScreenSize();
      Dimension frameSize  = frame.getSize();

      if (frameSize.width  > screenSize.width )
          frameSize.width  = screenSize.width;
      if (frameSize.height > screenSize.height)
          frameSize.height = screenSize.height;

      frame.setLocation (
        (screenSize.width  - frameSize.width ) >> 1,
        (screenSize.height - frameSize.height) >> 1
      );
   }
}

import java.awt.*;
import net.java.games.jogl.*;

public class PlanesView implements GLEventListener {

   /**
    * Take care of initialization here.
    */
   public void init(GLDrawable gld) {
      //The mode is GL.GL_MODELVIEW by default
      //We will also use the default ViewPort
      GL  gl  = gld.getGL();
      GLU glu = gld.getGLU();
```

```
  gl.glClearColor(0.0f, 0.0f, 0.0f, 1.0f);

  //Define points for eye, at and up.
  //This is your camera. It ALWAYS goes
  //in the GL_MODELVIEW matrix.
  glu.gluLookAt(
              20, 18,  0,
              20, 18, 30,
               0,  1,  0
              );

}

/**
 * Take care of drawing here.
 */
public void display(GLDrawable gld) {

  float red = 1.0f;
  float green = 0.0f;
  float blue = 0.5f;

  GL gl = gld.getGL();
  GLU glu = gld.getGLU();

  //We're changing the mode to GL.GL_PROJECTION
  //the only JOGL methods that should be called
  //while using the GL_PROJECTION matrix are:
  //    gluPerspective
  //    glFrustum
  //    glOrtho
  //    gluOrtho2D
  //    glLoadIdentity
  //    glLoadMatrix
  gl.glMatrixMode(GL.GL_PROJECTION);
  gl.glLoadIdentity();

  // Aspect is width/height
  double w = ((Component) gld).getWidth();
  double h = ((Component) gld).getHeight();
  double aspect = w/h;
  //Notice we're not using gluOrtho2D.
  //When using gluPerspective near and far need
  //to be positive.
```

```java
    //The arguments are:
    //fovy, aspect, near, far
    glu.gluPerspective(60.0, aspect, 25.0, 55.0);

    gl.glMatrixMode(GL.GL_MODELVIEW);
    //we don't want to initialize the GL_MODELVIEW
    //using gl.glLoadIdentity() this time. It has
    //settings from the init() that we wish to keep.

    gl.glClear(GL.GL_COLOR_BUFFER_BIT);

    gl.glColor3f(red, green, blue);
    //notice that the three squares are
    //exactly the same size. They appear
    //different on screen because of
    //perspective
    gl.glBegin(GL.GL_QUADS);

      //1st Plane
      gl.glVertex3i(0, 30, 30);
      gl.glVertex3i(10, 30, 30);
      gl.glVertex3i(10, 20, 30);
      gl.glVertex3i(0, 20, 30);

      //2nd Plane
      gl.glVertex3i(20, 20, 37);
      gl.glVertex3i(30, 20, 37);
      gl.glVertex3i(30, 10, 37);
      gl.glVertex3i(20, 10, 37);

      //3rd Plane
      gl.glVertex3i(40, 10, 45);
      gl.glVertex3i(50, 10, 45);
      gl.glVertex3i(50, 0, 45);
      gl.glVertex3i(40, 0, 45);

    gl.glEnd();
}

public void reshape(
                    GLDrawable drawable,
                    int x,
                    int y,
                    int width,
                    int height
```

```
                        ) {}

    public void displayChanged(
                        GLDrawable drawable,
                        boolean modeChanged,
                        boolean deviceChanged
                    ) {}
}
```

Chapter Six
Drawing Geometric Primitives

GL, GLU and GLUT

We've covered a lot of JOGL concepts so far, and I'm sure that you've already written an app or two. You have likely experimented with drawing different custom shapes using JOGL even though we haven't discussed drawing specifically.

You are about to be amazed at the number of drawing libraries and pre-made shapes available in JOGL. These are just some of the simple ones. We're not going to examine more advanced topics such as nurbs, vertex arrays, bitmaps or fonts. Frankly, we need to start somewhere and this book is intended to be an introductory tutorial.

It's time to explain a little about the C underpinnings of JOGL. JOGL uses JNI to communicate with the C based OpenGL libraries built into your graphics card or built into your Operating System (OS). These libraries, as mentioned above, are C based. The libraries are organized into header files. Three that directly impact this chapter are 'gl.h', 'glu.h' and 'glut.h'.

'gl.h' is the main header that must be used by every OpenGL program. 'glu.h' is used in most OpenGL programs too, but can be avoided sometimes if the program is very simplistic. GLU stands for OpenGL Utility Library.

Gene Davis

Try using that as a pickup line next time you're out clubbing.

'gl.h' and 'glu.h' provide all kinds of useful functions. (Remember we're still in C land here.) Unlike Java, C does not provide any graphics or Windowing Toolkit libraries as part of the base APIs. In Java we have two toolkits. We have the AWT and Swing. C coders are envious, trust me on this one.

So let's do a little computing of our own here. In C you can't make windows with the standard libraries. OpenGL MUST be drawn in windows. Oops!

Now there are various ways around this conundrum in most Operating Systems, but this is far more effort than most people want to go to just to learn OpenGL. Mark Kilgard came to the rescue luckily. He came up with the GLUT library. It is not officially part of OpenGL, but usually they are both on systems that support OpenGL.

GLUT stands for OpenGL Utility Toolkit. It is a very simple windowing toolkit, allowing for window creation, and also mouse and keyboard event tracking. This portion of GLUT is not needed for JOGL programmers since we have the AWT and Swing.

JOGL also contains GLUT functions to allow for cool pre-made shapes. If you want to check out the available methods, check the javadocs for net.java.games.jogl.util.GLUT.

GL 2D and 3D Primitives

Way back at the beginning of the book, and in examples throughout the book, we've used the methods glBegin() and glEnd() to enclose drawing instructions. These two commands surround glVertex commands. They must be placed at the beginning and end of the glVertex commands for the glVertex commands to be interpreted properly.

GL's glBegin() method is called first. It takes one of several static final values from the GL class as an argument. Below I've shown diagrams of the drawing styles for connecting (or not connecting) the vertices specified between glBegin() and glEnd(). Think of drawing with glVertex as making a kids dot-to-dot for the computer to play with.

For clarity, I've added numbers to some of the drawing patterns indicating the order each point was drawn in.

Learning Java Bindings for OpenGL (JOGL)

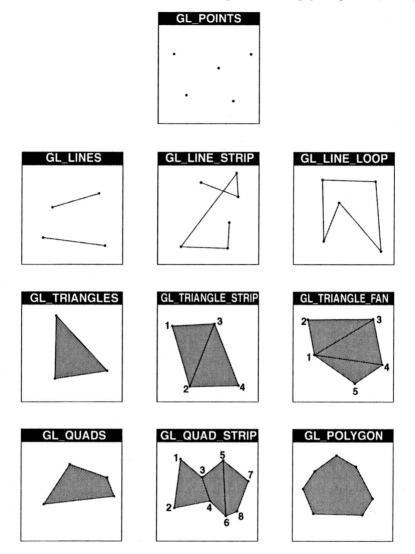

There are several versions of glVertex. The four that you will find most useful are glVertex2d(double x, double y), glVertex3d(double x, double y, double z), glVertex2i(int x, int y) and glVertex3i(int x, int y, int z). The glVertex2d and glVertex2i() methods are used for 2D drawings. This should be obvious, but on the off chance it wasn't I'm mentioning it.

Check the javadoc for the GL class for a full listing of glVertex*() methods. It's unlikely that you will need most of the calls now, but when your skills improve you may want to conserve memory or optimize code using some of the other glVertex calls that I haven't mentioned.

Notice two of the glVertex calls that I list take a 'z' coordinate. These allow for you to plot 3D arrangements of geometric shapes, lines or points.

One important note: OpenGL cannot draw concave polygons. Any shape (a quad or GL.GL_POLYGON) that you wish to draw that requires what appears to be a concave shape will need to be drawn by two or more convex shapes -- such as multiple quads or triangles -- creating the illusion of a concave shape. This is called subdividing. Below is an example.

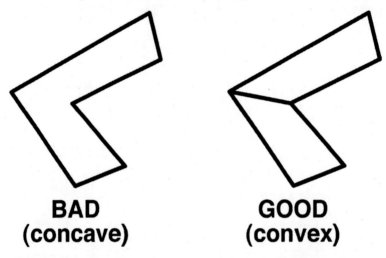

BAD (concave)   **GOOD (convex)**

It's easy to spot a concave shape. If any two corners in the shape cannot have a line drawn between them without that line leaving the shape, then the shape is not convex.

Many of the examples in this book have already shown how to use glBegin(), glVertex*(), and glEnd(). I see no point in beating a dead horse. Just flip back through previous chapters for examples on how to draw geometric primitives.

GLU Primitives

All the geometric primitives just drawn require a lot of planning. If you just want to create a shape that is 3D and not put in much work, the GLU class has that covered. Simple 3D shapes work well as place holders for more complex shapes that you haven't had time to design yet.

The GLU class contains method calls for a few nifty pieces of eye candy. Methods allow you to create cylinders, spheres and disks. This is done using quadrics.

You remember that drawing the geometric primitives needed three different calls: You had a glBegin() call, followed by a group of glVertex*() calls, and then you finished up with a call to glEnd().

This pattern is similar to that used by quadrics. For instance, here is the code for creating a sphere.

```
GLUquadric quad = glu.gluNewQuadric();
glu.gluQuadricDrawStyle(quad, GLU.GLU_LINE);
glu.gluSphere(quad, 1.0, 15, 15);
glu.gluDeleteQuadric(quad);
```

First we create a GLUquadric object. GLUquadric is found in the 'net.java.games.jogl.GLUquadric' package. We called it 'quad'.

Next we use our GLU instance 'glu' to set the draw style. There are several draw styles. They are contained in the GLU class. The valid draw styles are: GLU.GLU_POINT, GLU.GLU_LINE, GLU.GLU_SILHOUETTE and GLU.GLU_FILL. Typically you will only use point, line and fill.

The third step is to create the sphere, cylinder or disk. We do this with the methods gluSphere(GLUquadric quad, double radius, int slices, int stacks), gluCylinder(GLUquadric quad, double base, double top, double height, int slices, int stacks) or gluDisk(GLUquadric quad, double inner, double outer, int slices, int loops). We'll look at these methods in more detail in a moment.

The last step is to destroy the quadric with a call to gluDeleteQuadric(). Notice you passed in a GLUquadric to create a sphere, cylinder or disk. Once delete has been called for the quadric, you may no longer use that instance for drawing.

The gluSphere() method takes parameters of a GLUquadric, a radius, slices and stacks. The GLUquadric is the quadric we created in our first step. The radius is, well the size we want the radius in relation to the coordinate system. Think of slices and stacks in terms of resolution. The more we have, the better looking the sphere can be. More slices and stacks can affect rendering speed in some situations. Slices are like the longitude lines on a globe. Stacks are like the latitude lines on a globe.

The gluCylinder() method works just like the gluSphere except that there are two radiuses. There is a top radius and a bottom radius.

The gluDisk() method creates a disk, like a CD or LP (if you're really old ;-). The inner and outer parameters represent the outer edge and the inner hole in the disk. Zero may be entered for the inner radius if you don't want a hole in the disk. Slices are just that. (How many slices do you want to cut this pie into anyway?) Rings are the number of rings around the disk.

So the big question that is probably going through your mind at this moment is, "Where is this thing getting drawn?" It's true, no parameters are given to describe where the sphere, cylinder or disk are drawn; just their sizes.

The answer is centered on the origin. In the next chapter we'll explain how to move them about. For now place the camera looking at the origin and make sure the entire object fits in the perspective we've chosen.

The example for drawing GLUT primitive in the next section has some commented code for creating a sphere. Comment the GLUT shape and uncomment the GLU shape code to see a GLU sphere in action.

GLUT Primitives

The GLUT class only provides a subset of the functions GLUT provides in C. Still it manages to one-up GLU for quick and simple 3D shapes. At the time of this writing, the GLUT class provides only methods for creating shapes and text.

The GLUT class can be found in the package 'net.java.games.jogl.util'. It provides methods for creating cubes, cones, bitmapped strings, dodecahedrons, spheres and much more.

There are only two steps to drawing GLUT shapes. First create a GLUT object as you would any Java object. Then using a valid GL instance or GLU instance call the desired method on the newly created GLUT object. Nothing fancy there.

These objects are created at the origin of the current coordinate system.

Learning Java Bindings for OpenGL (JOGL)

```
import java.awt.*;
import java.awt.event.*;
import javax.swing.*;
import net.java.games.jogl.*;

/**
 * This is a basic JOGL app. Feel free to
 * reuse this code or modify it.
 */
public class GlutObjects extends JFrame {

  public static void main(String[] args) {
    final GlutObjects app = new GlutObjects();

    //show what we've done
    SwingUtilities.invokeLater (
      new Runnable() {
        public void run() {
          app.setVisible(true);
        }
      }
    );
  }

  public GlutObjects() {
    //set the JFrame title
    super("Glut Objects Application");

    //kill the process when the JFrame is closed
```

```java
        setDefaultCloseOperation(JFrame.EXIT_ON_CLOSE);

        //only three JOGL lines of code ... and here they are
        GLCapabilities glcaps = new GLCapabilities();
        GLCanvas glcanvas = GLDrawableFactory.getFactory().createGLCanvas(glcaps);
        glcanvas.addGLEventListener(new GlutObjectsView());

        //add the GLCanvas just like we would any Component
        getContentPane().add(glcanvas, BorderLayout.CENTER);
        setSize(500, 300);

        //center the JFrame on the screen
        centerWindow(this);
    }

    public void centerWindow(Component frame) {
        Dimension screenSize = Toolkit.getDefaultToolkit().getScreenSize();
        Dimension frameSize = frame.getSize();

        if (frameSize.width > screenSize.width ) frameSize.width  = screenSize.width;
        if (frameSize.height > screenSize.height) frameSize.height = screenSize.height;

        frame.setLocation (
           (screenSize.width  - frameSize.width ) >> 1,
           (screenSize.height - frameSize.height) >> 1
        );
    }
}

import java.awt.*;
import net.java.games.jogl.*;
//We need a new package for GLUT shapes
import net.java.games.jogl.util.*;

public class GlutObjectsView implements GLEventListener {
```

```java
/**
 * Take care of initialization here.
 */
public void init(GLDrawable gld) {
  //The mode is GL.GL_MODELVIEW by default
  //We will also use the default ViewPort
  GL gl = gld.getGL();
  GLU glu = gld.getGLU();

   //We're changing the mode to GL.GL_PROJECTION
   //This is where we set up the camera
   gl.glMatrixMode(GL.GL_PROJECTION);
   gl.glLoadIdentity();

  // Aspect is width/height
  double w = ((Component) gld).getWidth();
  double h = ((Component) gld).getHeight();
  double aspect = w/h;
  //When using gluPerspective near and far need
  //to be positive.
  //The arguments are:
  //fovy, aspect, near, far
  glu.gluPerspective(60.0, aspect, 2.0, 13.0);
}

/**
 * Take care of drawing here.
 */
public void display(GLDrawable gld) {

  float red   = 0.0f;
  float green = 0.0f;
  float blue  = 0.9f;

  GL gl = gld.getGL();
  GLU glu = gld.getGLU();

  gl.glMatrixMode(GL.GL_MODELVIEW);
  gl.glLoadIdentity();

  gl.glClearColor(0.0f, 0.0f, 0.0f, 1.0f);

  //Define points for eye, at and up.
  //This is your camera. It ALWAYS goes
```

```
            //in the GL_MODELVIEW matrix.
            glu.gluLookAt(
                        0, 0, 4,
                        0, 0, 0,
                        0, 1, 0
                        );

            gl.glClear(GL.GL_COLOR_BUFFER_BIT);

            gl.glColor3f(red, green, blue);

            //We use wire here because default
            //lighting is not good enough to
            //use when rendering the solid version
            GLUT glut = new GLUT();
            glut.glutWireIcosahedron(gl);

            //The two lines above could be replaced
            //with the glu library code that follows.
            //This produces a line style sphere.

            //GLUquadric quad = glu.gluNewQuadric();
            //glu.gluQuadricDrawStyle(quad, GLU.GLU_LINE);
            //glu.gluSphere(quad,  1.0,  15,  15);
            //glu.gluDeleteQuadric(quad);

    }

    public void reshape(
                    GLDrawable drawable,
                    int x,
                    int y,
                    int width,
                    int height
                    ) {}


    public void displayChanged(
                    GLDrawable drawable,
                    boolean modeChanged,
                    boolean deviceChanged
                    ) {}
}
```

Chapter Seven
First Person Movement in 3D Space

Transformations

We can move around all we want, but if we have nothing to use as a point of reference, we won't feel like we are moving. Let's use some shapes as our reference, but to do this we will need to discuss transformations. These transformations will happen to the GL_MODELVIEW.

The pre-made shapes we have used so far are all positioned at the origin when they are created. This is fine for one object, but not very useful if we want more than one object, or if we don't want them on the origin. They are always the same size and facing the same direction.

There are three types of transformations in JOGL: Scale, performed by GL's glScaled(double x, double y, double z) method; rotate, performed by the glRotated(double angle, double x, double y, double z) method; and translate, performed by the glTranslated(double x, double y, double z) method.

Mainly we are interested in moving these objects away from the origin, though. So I'll focus on the glTranslate method. After you feel comfortable with translating, you will want to learn to scale and rotate too. Don't be afraid to experiment with them.

There are actually two methods for translating, glTranslated() and glTranslatef(). These are the C function names. If JOGL had been written before OpenGL, these two methods would have just been overloaded so that they had the same name. Alas, not so.

glTranslated() takes doubles as arguments. glTranslatef() takes floats as arguments. The reason for this was to work around C's inability to overload method names.

With glTranslated(), you still draw objects at the origin, kind of. Think of the origin as a soccer ball. Now pretend you have a strange fetish for drawing a square in the dirt around that ball.

You see this is going to get boring very quickly. You need variety in your obsessions. So, you kick the ball out of frustration. You quickly realize that you can draw a new square around the ball in dirt that you've never drawn in before. You can keep kicking the ball and finding a new place to draw your squares.

The origin is the soccer ball. glTranslated() is kicking the soccer ball. Remember, you need to kick the ball before you can draw somewhere new.

If you have an object that is always drawn at the origin, but you want it somewhere else, move the origin by passing an x, y, and z value to modify the origin by to glTranslated(). Always call glTranslated() before drawing the items you wish to drawn with the translation.

Describing the Example

This sample is not too complex, but let's spend a little effort describing how this sample works. First we needed a KeyListener. I've modified the KeyListener example from our chapter on Events and Listeners. If you're not familiar with KeyListeners, review the section on them in our earlier discussion.

The FirstPersonView class is acting as both our GLEventListener and our KeyListener. For readability I pass an instance of GLCanvas into FirstPersonView. You can obtain the GLCanvas from the KeyEvent though, and this may be preferable in many instances. To repaint using the KeyEvent's instance of the GLCanvas make a call similar to 'theKeyEvent.getComponent().repaint();'. You need the GLCanvas instance for repaint() calls.

The keys used for moving around in the space with our shapes are '8', '6', '2' and '4'. '8' is forward, '2' is backward, '4' is left and '6' is right.

Each time we type one of the four movement keys, we change the position of the camera. It is set in the gluLookAt() method. After positioning the camera in the default coordinate system, we start messing with creating shapes and translating the coordinate system. I've made the shapes several different colors to provide a little more variety.

Pretty much the only thing that we do in the GLEventListener's init() method is set up the camera using the gluPerspective() method. All the work of positioning the camera and drawing the shapes is done in the display() method.

The Code

"Words, words, words" ... let's get to the code.

```
import java.awt.*;
import java.awt.event.*;
import javax.swing.*;
import net.java.games.jogl.*;

/**
 * This is a basic JOGL app. Feel free to
 * reuse this code or modify it.
 */
public class FirstPersonMovement extends JFrame {

  static GLCanvas glcanvas = null;

  public static void main(String[] args) {
    final FirstPersonMovement app = new FirstPersonMovement();

    // show what we've done
    SwingUtilities.invokeLater (
      new Runnable() {
        public void run() {
          app.setVisible(true);
          glcanvas.requestFocusInWindow();
        }
      }
    );
  }

  public FirstPersonMovement() {
    //set the JFrame title
    super("First Person Movement");

    //kill the process when the JFrame is closed
```

```java
    setDefaultCloseOperation(JFrame.EXIT_ON_CLOSE);

    //create our FirstPersonView which serves two purposes
    // 1) it is our GLEventListener, and
    // 2) it is our KeyListener
    FirstPersonView fpv = new FirstPersonView();

    //only three JOGL lines of code ... and here they are
    GLCapabilities glcaps = new GLCapabilities();
    glcanvas =
        GLDrawableFactory.getFactory().createGLCanvas(glcaps);
    glcanvas.addGLEventListener(fpv);
    glcanvas.addKeyListener(fpv);

    //we'll want this for our repaint requests
    fpv.setGLCanvas(glcanvas);

    //add the GLCanvas just like we would any Component
    getContentPane().add(glcanvas, BorderLayout.CENTER);
    setSize(500, 300);

    //center the JFrame on the screen
    centerWindow(this);
  }

  public void centerWindow(Component frame) {
    Dimension screenSize =
        Toolkit.getDefaultToolkit().getScreenSize();
    Dimension frameSize = frame.getSize();

    if (frameSize.width > screenSize.width )
       frameSize.width = screenSize.width;
    if (frameSize.height > screenSize.height)
       frameSize.height = screenSize.height;

    frame.setLocation (
      (screenSize.width - frameSize.width ) >> 1,
      (screenSize.height - frameSize.height) >> 1
    );
  }
```

}

Now for the fun class.

```java
import java.awt.*;
import java.awt.event.*;
import net.java.games.jogl.*;
import net.java.games.jogl.util.*;

/**
 * For our purposes only two of the
 * GLEventListeners matter. Those would
 * be init() and display().
 */
public class FirstPersonView

    implements GLEventListener, KeyListener
{

  int xPosition = 0;
  int zPosition = 0;

  float red = 0.0f;
  float green = 0.0f;
  float blue = 1.0f;

  GLCanvas glc;

  public void setGLCanvas(GLCanvas glc) {
    this.glc = glc;
  }

  /**
   * Take care of initialization here.
   */
  public void init(GLDrawable drawable) {
    //The mode is GL.GL_MODELVIEW by default
    //We will also use the default ViewPort
    GL gl = drawable.getGL();
    GLU glu = drawable.getGLU();

    //We're changing the mode to GL.GL_PROJECTION
    //This is where we set up the camera
    gl.glMatrixMode(GL.GL_PROJECTION);
    gl.glLoadIdentity();
```

```
  // Aspect is width/height
  double w = ((Component) drawable).getWidth();
  double h = ((Component) drawable).getHeight();
  double aspect = w/h;
  //When using gluPerspective near and far need
  //to be positive.
  //The arguments are:
  //fovy, aspect, near, far
  glu.gluPerspective(60.0, aspect, 2.0, 20.0);
}

/**
 * Take care of drawing here.
 */
public void display(GLDrawable drawable) {

  GL gl = drawable.getGL();
  GLU glu = drawable.getGLU();
  GLUT glut = new GLUT();

  gl.glMatrixMode(GL.GL_MODELVIEW);
  gl.glLoadIdentity();

  gl.glClearColor(0.0f, 0.0f, 0.0f, 1.0f);

  //Define points for eye, at and up.
  //This is your camera. It ALWAYS goes
  //in the GL_MODELVIEW matrix.
  glu.gluLookAt(
              xPosition, 0, zPosition,
              xPosition, 0, (zPosition+20),
              0, 1, 0
              );

  gl.glClear(GL.GL_COLOR_BUFFER_BIT);


  red   = 0.0f;
  green = 0.0f;
  blue  = 0.9f;

  gl.glColor3f(red, green, blue);

  //transforming the place the next shape
```

```
//will be drawn.
gl.glTranslated(2, 0, 2);

//We use wire here because default
//lighting is not good enough to
//use when rendering the solid version
glut.glutWireIcosahedron(gl);

//more shapes to navigate through
gl.glTranslated(-4, 0, 0);
glut.glutWireIcosahedron(gl);

red   = 0.0f;
green = 0.9f;
blue  = 0.1f;
gl.glColor3f(red, green, blue);
gl.glTranslated(4, 0, 4);
glut.glutWireIcosahedron(gl);
gl.glTranslated(-4, 0, 0);
glut.glutWireIcosahedron(gl);

red   = 0.9f;
green = 0.0f;
blue  = 0.1f;
gl.glColor3f(red, green, blue);
gl.glTranslated(4, 0, 4);
glut.glutWireIcosahedron(gl);
gl.glTranslated(-4, 0, 0);
glut.glutWireIcosahedron(gl);

red   = 0.9f;
green = 0.0f;
blue  = 0.9f;
gl.glColor3f(red, green, blue);
gl.glTranslated(4, 0, 4);
glut.glutWireIcosahedron(gl);
gl.glTranslated(-4, 0, 0);
glut.glutWireIcosahedron(gl);

red   = 0.5f;
green = 0.5f;
blue  = 0.5f;
gl.glColor3f(red, green, blue);
gl.glTranslated(4, 0, 4);
glut.glutWireIcosahedron(gl);
```

```
    gl.glTranslated(-4, 0, 0);
    glut.glutWireIcosahedron(gl);



}

public void reshape(
                    GLDrawable drawable,
                    int x,
                    int y,
                    int width,
                    int height
                ) {}

public void displayChanged(
                    GLDrawable drawable,
                    boolean modeChanged,
                    boolean deviceChanged
                ) {}

//////////////////////////////////
// KeyListener implementation below

public void keyPressed(KeyEvent e) {}

public void keyReleased(KeyEvent e) {}

public void keyTyped(KeyEvent e) {
    if (e.getKeyChar() == KeyEvent.VK_8)
        zPosition += 1;
    else if (e.getKeyChar() == KeyEvent.VK_2)
        zPosition -= 1;
    else if (e.getKeyChar() == KeyEvent.VK_4)
        xPosition += 1;
    else if (e.getKeyChar() == KeyEvent.VK_6)
        xPosition -= 1;

    glc.repaint();
  }
}
```

All this should produce a window you can move around in. Once again, the movement keys are '4', '8', '6' and '2'. Using the number pad on your keyboard will make it easy.

Here is the view.

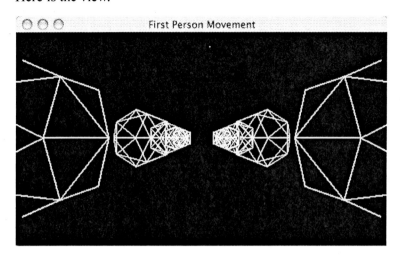

Gene Davis

Chapter Eight
Lights

Enabling Lighting

So far we have not made any 3D solid shapes. If you have tried, then you probably ended up with a silhouette of the shape and no visible details on the shape. The problem is that every surface on the shape is the same color. The color has been set by a call to glColor3f() in our GL instance.

In the real world no object is one color on all sides. If it was we would not perceive depth as readily. If you track down an object such as a block or ball that is all one color and really take a hard look at it you'll notice that the surface varies in shading and may even be reflecting some of the colors of objects around it. Even the color of the sunlight or light bulb illuminating its surface changes an object's viewed color.

We need a light source in our scenes if we want this kind of real world depth and contrast of color and shading. We don't get a light source automatically. We have to jump though a few hoops first. We need to tell the OpenGL machine that we want lighting. We do this with a call to GL's glEnable(GL.GL_LIGHTING) method. All that call does is say that we want to provide our own light source.

As soon as we've enabled GL_LIGHTING, our calls to glColor3f() take on a different meaning and to us are useless. All our objects will be grey by default. We'll fix this using "materials" later on.

Enabling the GL_LIGHTING alone does not prepare us to enable the light. We'll need to set the position of the light and the amount of light coming from it, but we'll get back to that.

Learning Java Bindings for OpenGL (JOGL)

We are given a fixed set of disabled lights to use. The number of lights we have access to depends on the implementation of OpenGL that JOGL is interfacing with. You will have at least eight, but the actual number is defined in GL.GL_MAX_LIGHTS. The lights are associated with several of the GL class's static attributes. They are named GL_LIGHT0, GL_LIGHT1, GL_LIGHT2, GL_LIGHT3, (etc.) This convention may seem very strange to a Java programmer, but it works well in C, which is what JOGL interfaces with to use OpenGL.

Lights are turned on by yet another call to the glEnable() method using the static variable representing the light we wish to use. For example 'gl.glEnable(GL.GL_LIGHT0); glEnable() gets used all over the place, so you'll be seeing a lot of it in your JOGL programs.

Diffuse and Ambient Light

For our purposes, lights produce two kinds of light. (Actually there are more, but we're keeping this basic.) The two kinds of light are "diffuse" and "ambient." The diffuse property sets the color and brightness of our light's emitted light.

To set up the diffuse property for the light we create a float array with four values. The first three are the important values for now. They represent the red, green and blue values. Set them to values between 0.0f and 1.0f. Zero to one is your default scale, but you can change your scale by setting one of the values to something greater than one. In that case the highest value is considered to be the equivalent of 1.0f (the brightest value), and the other values are scaled appropriately.

The fourth value is set to zero. You'll modify it when your JOGL skills are more advanced.

The float array representing diffuse lighting is then input using a call similar to this:

```
gl.glLightfv(GL.GL_LIGHT0, GL.GL_DIFFUSE, diffuse);
```

As I mentioned earlier, the diffuse light is the regular light coming from our GL_LIGHT0. Ambient light is the amount of light that seems to have no direction or source that comes originally from our light source.

For example, let's say that we have a regular yellow light bulb in a room with no furniture. Imagine every wall, the floor and ceiling are painted red. If you were to read a book that is under the light it would look pretty normal. If you held the book between the light bulb and your eyes you would still be able to see the pages even though the book's pages are not directly exposed to the light bulb's light.

The light that is illuminating the pages in this instance is ambient light. Ambient light is light that has bounce off of one or more objects and seems to come from everywhere.

Ambient light usually has a color similar to the light sources, but that may not always be the case. For instance in our all red room the ambient light would take on a red hue depending on how reflective the paint is.

The ambient property is set pretty much the same way that the diffuse property is set. Create a float array with four elements. The fourth element will be zero. The first three elements represent RGB colors. The colors are scaled from 0.0f to 1.0f. The actual code will look like this:

```
float ambient[] = {.2f, .2f, .2f, 0f};
gl.glLightfv(GL.GL_LIGHT0, GL.GL_AMBIENT, ambient);
```

Positioning the Light

I've shown you how to turn the light on and set the diffuse and ambient colors for the emitted light. Let's set its position. Setting the position looks exactly like the code for setting colors for the light. The first three float values are the x, y and z coordinates. Just set the fourth element of the float array to zero for now. The code for setting the position looks like this:

```
float position[] = {0f, 15f, -30f, 0f};
gl.glLightfv(GL.GL_LIGHT0, GL.GL_POSITION, position);
```

You will notice that the glLightfv() method's first argument is GL.GL_LIGHT0 in our examples. This could have specified the set up for one of our other lights if we were enabling others. It could have just as easily been GL.GL_LIGHT5 or GL.GL_LIGHT7.

Depth Test

JOGL needs to be warned that some of what it draws should not actually be seen. If a polygon exists in the clipping area that should be obscured by another polygon it will be drawn anyway. The GLCanvas will gladly display whatever is drawable in the order it is found regardless of what should be in front. This is done because it is quick.

In some optimized systems, the engine is never passed anything that shouldn't be drawn. We, however, are not worried about performance and need to warn JOGL of our intention to give it polygons that should be ignored or partially obscured.

We warn JOGL by calling 'gl.glEnable(GL.GL_DEPTH_TEST)'. We also clear the depth buffer bit by tacking '| GL.GL_DEPTH_BUFFER_BIT' on to the glClear command that we've been using to set the background color of our scenes.

If we don't make sure that we're testing the depth of every thing in our scenes, we will end up with objects that look inside out or objects that look wrong in random ways.

A warning is in order. Remember the gluPerspective() call which we use to set up the virtual camera that we use with our scene. If the 'near' argument is too close to zero, then the depth test will appear as though it is not working. I would recommend never setting 'near' to less than '0.1'. This will prevent most problems.

Materials

Everything is grey by default. Unless you really like grey, you won't be satisfied with this. To make this easy we'll always use two parameters GL.GL_FRONT_AND_BACK and GL.GL_AMBIENT_AND DIFFUSE. This will make life easier for you. As you learn more about JOGL you'll change these values, but for now that information would just bog you down.

The color of objects depends on the lighting and the materials that are associated with the object. Material color is set with a four element float array. The first three elements represent the standard RGB colors. Just leave the last element at 1.0f for now. The actual call that sets the color is glMaterialfv().

This is what the code will look like:

```
float material[] = {0.9f, 0.6f, 0.06f, 1.0f};
gl.glMaterialfv(
    GL.GL_FRONT_AND_BACK,
    GL.GL_AMBIENT_AND_DIFFUSE,
    material
);
```

The current material will apply to each object drawn until it is changed with another call to glMaterialfv().

Gene Davis

Putting It All Together

Here is an example showing all of what I've explained in this chapter. I suggest you play with the lighting and material settings just to get a feel for what can be done.

```
import java.awt.*;
import javax.swing.*;
import net.java.games.jogl.*;

/**
 * This is a basic JOGL app. Feel free to
 * reuse this code or modify it.
 */
public class LightingApp extends JFrame {

  static GLCanvas glcanvas = null;

  public static void main(String[] args) {
    final LightingApp app = new LightingApp();

    // show what we've done
    SwingUtilities.invokeLater (
      new Runnable() {
        public void run() {
          app.setVisible(true);
          glcanvas.requestFocusInWindow();
        }
      }
```

```
    );
  }

  public LightingApp() {
    //set the JFrame title
    super("Lighting Example");

    //kill the process when the JFrame is closed
    setDefaultCloseOperation(JFrame.EXIT_ON_CLOSE);

    //create our FirstPersonView which serves two purposes
    // 1) it is our GLEventListener, and
    // 2) it is our KeyListener
    LightingView lv = new LightingView();

    //only three JOGL lines of code ... and here they are
    GLCapabilities glcaps = new GLCapabilities();
    glcanvas =
        GLDrawableFactory.getFactory().createGLCanvas(glcaps);
    glcanvas.addGLEventListener(lv);

    //add the GLCanvas just like we would any Component
    getContentPane().add(glcanvas, BorderLayout.CENTER);
    setSize(500, 300);

    //center the JFrame on the screen
    centerWindow(this);
  }

  public void centerWindow(Component frame) {
    Dimension screenSize =
        Toolkit.getDefaultToolkit().getScreenSize();
    Dimension frameSize  = frame.getSize();

    if (frameSize.width  > screenSize.width )
       frameSize.width  = screenSize.width;
    if (frameSize.height > screenSize.height)
       frameSize.height = screenSize.height;

    frame.setLocation (
```

Gene Davis

```
      (screenSize.width  - frameSize.width ) >> 1,
      (screenSize.height - frameSize.height) >> 1
    );
  }
}

import java.awt.*;
import net.java.games.jogl.*;
import net.java.games.jogl.util.*;

/**
 * For this program only one of the
 * GLEventListener's methods matter. It
 * would be display().
 */
public class LightingView implements GLEventListener {

  public void init(GLDrawable drawable) {}

  /**
   * Take care of drawing here.
   */
  public void display(GLDrawable drawable) {
    GL gl = drawable.getGL();
    GLU glu = drawable.getGLU();
    GLUT glut = new GLUT();


    //We're changing the mode to GL.GL_PROJECTION
    //This is where we set up the camera
    gl.glMatrixMode(GL.GL_PROJECTION);
    gl.glLoadIdentity();

    // Aspect is width/height
    double w = ((Component) drawable).getWidth();
    double h = ((Component) drawable).getHeight();
    double aspect = w/h;
    //When using gluPerspective near and far need
    //to be positive.
    //The arguments are:
    //fovy, aspect, near, far
    glu.gluPerspective(60.0, aspect, 2.0, 10.0);

    gl.glMatrixMode(GL.GL_MODELVIEW);
    gl.glLoadIdentity();
```

```
    gl.glClearColor(0.0f, 0.0f, 0.0f, 1.0f);

    //Define points for 'eye', 'at' and 'up'
    //This is your camera. It ALWAYS goes
    //in the GL_MODELVIEW matrix.
    glu.gluLookAt(
                0, 0, 0,
                0, 0, 20,
                0, 1, 0
                );

    //Time to set up the light. We will only
    //use one light. If we used multiple lights,
    //we would want to dim them so as to not
    //bleach out the scene with too much light.
    float position[] = {0f, 15f, -30f, 0f};
    gl.glLightfv(GL.GL_LIGHT0, GL.GL_POSITION,
position);

    float diffuse[] = {.7f, .7f, .7f, 0f};
    gl.glLightfv(GL.GL_LIGHT0, GL.GL_DIFFUSE,
diffuse);

    float ambient[] = {.2f, .2f, .2f, 0f};
    gl.glLightfv(GL.GL_LIGHT0, GL.GL_AMBIENT,
ambient);

    gl.glEnable(GL.GL_LIGHTING);
    gl.glEnable(GL.GL_LIGHT0);
    //GL_DEPTH_TEST prevents us from seeing polygons
    //that should be obscured. Remember to clear the
    //GL.GL_DEPTH_BUFFER_BIT
    gl.glEnable(GL.GL_DEPTH_TEST);

    //Notice the depth buffer bit is also being
cleared
    gl.glClear(
       GL.GL_COLOR_BUFFER_BIT |
       GL.GL_DEPTH_BUFFER_BIT
    );


    //Placing icosahedron
    //The icosahedron takes a GL instance
```

```
//as a parameter.
gl.glTranslated(0, -.3, 5);
float material1[] = {0.9f, 0.6f, 0.06f, 1.0f};
gl.glMaterialfv(
   GL.GL_FRONT_AND_BACK,
   GL.GL_AMBIENT_AND_DIFFUSE,
   material1
);
glut.glutSolidIcosahedron(gl);

//Placing cube
//The cube takes a GL instance and
//a edge length as arguments.
gl.glTranslated(3, 0, 0);
float material2[] = {0.9f, 0f, 0f, 1.0f};
gl.glMaterialfv(
   GL.GL_FRONT_AND_BACK,
   GL.GL_AMBIENT_AND_DIFFUSE,
   material2
);
glut.glutSolidCube(gl, 1.2f);

//Placing sphere
//The Cone parameters are:
//   GLU instance (NOT GL)
//   radius of base
//   height of cone
//   slices or subdivisions around cone
//   stacks or subdivisions from base to peak
gl.glTranslated(-6, 0, 0);
float material3[] = {0.05f, 0f, 0.7f, 1.0f};
gl.glMaterialfv(
   GL.GL_FRONT_AND_BACK,
   GL.GL_AMBIENT_AND_DIFFUSE,
   material3
);
glut.glutSolidSphere(glu, .8, 100, 360);
}

public void reshape(
                  GLDrawable drawable,
                  int x,
                  int y,
                  int width,
```

```
                        int height
                    ) {}

    public void displayChanged(
                        GLDrawable drawable,
                        boolean modeChanged,
                        boolean deviceChanged
                    ) {}
}
```

Chapter Nine
Textures

Why Textures?

I'll start by warning you that this is the toughest chapter to follow in the book. Don't stress it. A lot of what I do here, you can mimic until you understand it enough to do your own thing. I'm going to mention a lot of method calls without giving too many details. The purpose for this is to not bog you down with details that you don't need at this point. I will give you enough information to aid further research on topics when you feel you are ready for them.

Having said that, "Why use textures?" I'd answer, "Speed and looks, that's why." Most JOGL programs will use them to speed rendering of scenes and make them fancier. Besides, ... all the cool kids are doing it!

Making Textures

One topic books always seem to gloss over is that textures don't just pop into existence. Someone is making them, and they sure are expensive to buy. Let's spill the beans, shall we? Here's how to make textures.

Textures are made from images. The main caveat is that the image dimensions must measure a number that is a power of two. By powers of two I mean numbers like 1, 2, 4, 8, 16, 32, 64, ... (etc.) So an image that is 4 x 16 would work just fine, but an image with dimensions of 3 x 16 wouldn't work, because 3 is not a power of 2.

Learning Java Bindings for OpenGL (JOGL)

Any type of image may be used. You will have to experiment or do some research when you want to load it and convert it into a form usable by JOGL. It would be nice to see some methods to automate the process. I haven't heard of any common ones yet.

I will convert my image to an eight bit per channel RGB image and save it as a PNG. You will need a powerful photo manipulation program like PhotoShop. Gimp would probably work, but PhotoShop is more widely used so I'll do the conversion using it.

Here is what the original picture looks like. It was taken with a cheap digital camera. You can pick any pattern in nature or man-made. Some patterns will be more difficult than others to prepare, but they'll be cheaper than buying texture CDs.

You have to make some decisions early on. Where will the light source come from in your scene? What color will that light be? Your initial picture should mimic your intended use. This will make the texture move believable.

Which part of the pattern do you like the most? Remember, any distinctive features will need to be removed if you wish to prevent the pattern repetition from detracting from your scene. It may not matter much though if it is a minor part of your program.

I chose this section and cropped the image to only include this.

I had to rotate the image slightly to line up the bricks. At this point I cleaned up the image a bit to remove features I didn't want. The clone tool is perfect for this.

Now I resampled the image. Resampling it I chose a width and size that were powers of 2. In this case it warped the image, but that can be fixed when the image is placed in the JOGL program. The image became 128 x 256.

Lastly I used the "offset" filter to wrap the image 128 pixels horizontally and 64 pixels vertically. Offsetting the image created this image (slightly pixellated by now). Notice the distinctive lines forming a plus through the center of the texture.

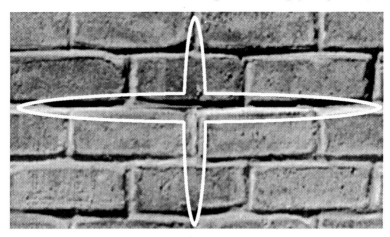

Remove the vertical and horizontal lines using the clone tool. Lastly, rotate the image if necessary and save the image in the desired format (PNG).

You're done. Best of all, you have no copyrights to be hassled with.

Using Textures

To use an image for a texture, you need to load it. I'll assume that the image you want to use is a PNG in your JAR file. I chose PNG because that format has few legal restrictions for use.

The PNG image I've used in this chapter is an RGB with eight bits per channel. Another type of PNG or image may or may not work with this sample code. A good photo manipulation program such as PhotoShop will allow you to check the format of any image you desire to use and even change the format of the image if you wish.

I'll detail the command to get the brick image into the JAR file later. For now, I'll focus on getting the image from the JAR file into the running program and preparing it for use as a texture.

I've separated the loading and preparing of the PNG file to be a texture into this call:

`mipmapsFromPNG(GLU glu, String imgName)`

One of the common problems presented to JOGL and OpenGL programmers is, "How do I make images usable by my program?" (Importing images are easy compared to importing 3D objects. Trust me.)

The mipmapsFromPNG() method makes it easy to use PNGs. There are quite a few methods freely available, but you'll have to do some digging through source code to find them. This one is pretty basic and inflexible (read: short and easy.)

The 'imgName' parameter is actually the path to the file in the JAR.

I've included many comments in the code. It would be frivolous to repeat it all here. Read the code (don't forget the comments) and mess around with the program. You'll learn the most by debugging your own experiments.

Another chunk of code I'd like to draw your attention to is:

```
// x magnification, x shear,
// z magnification, x translation
float sMapping[] = { 0.1f, 0f, 0f, 0f };

// y shear, y magnification,
// z magnification, y translation
float tMapping[] = { 0f, 0.2f, 0f, 0f };
```

Modifying these parameters will allow you to move (translate) the texture up, down, left and right. You may also stretch (magnify) the image vertically or horizontally. Shear allows you move one edge of the texture separately, providing a skewing effect.

Pop Goes the Matrix

Now we're going to take a slight detour. I don't want you reading this entire book and not learning how to push and pop a matrix. Think of pushing and popping as a way to save state.

If you have ever played a video game, chances are that you've saved it so that you could go eat dinner or watch TV. Some people will save the game after doing something hard, like finishing a level. Then, if they lose the game, they go back to the saved game and start from there. It's a handy "cheat".

The same thing is done when positioning objects in JOGL. You know you're about to move the origin off to Timbuktu. This will make positioning other objects hard. So, you save your location using a call to glPushMatrix(). Then, when you are finished making your object in Timbuktu, you start back where you were by calling glPopMatrix();

Let's look at this another way. If you were exploring a cave, you would leave a way to get back to the entrance so you could start over. When you are about to change the position that you are going to draw objects at in the GL_MODELVIEW matrix, you can save the position (say, the origin) you were at in memory by 'pushing' the matrix. You can do any transformations, such as glTranslated(), draw your object and then return to your initial position by 'popping' the GL_MODELVIEW.

Bricks the Program

Preparing this program is a bit more complex than any that we have done so far. Don't worry, be happy.

After compiling the two classes in this example, this program and its resources need to be JARed. The command is,

```
jar cvfm tex_app.jar Manifest *.class an_img.png
```

I have included both classes (actually three classes when compiled), an image and a manifest file. The image and other files will be in a new JAR file called 'tex_app.jar' now.

The manifest file is just a text file with one line -- 'Main-Class: TexturesApp'. The text file called 'Manifest' does not have an extension.

The image is any '*.png' that you wish to use as your texture in this example. The '*.png' I used will be available on my website (www.gene davissoftware.com) if you wish to use it. If you bought an electronic or traditional version of this book, feel free to use the texture in any program you write free of charge.

You can start this program by double-clicking the JAR file or by typing 'java -jar tex_app.jar' at the command line.

Gene Davis

Here is the code to compile. I've commented the 'TexturesApp.java' file extensively, so be sure to read the comments!

```java
import java.awt.*;
import javax.swing.*;
import net.java.games.jogl.*;

/**
 * This is a basic JOGL app. Feel free to
 * reuse this code or modify it.
 */
public class TexturesApp extends JFrame {

  static GLCanvas glcanvas = null;

  public static void main(String[] args) {
    final TexturesApp app = new TexturesApp();

    // show what we've done
    SwingUtilities.invokeLater (
      new Runnable() {
        public void run() {
          app.setVisible(true);
          glcanvas.requestFocusInWindow();
        }
      }
    );
  }

  public TexturesApp() {
    //set the JFrame title
    super("Textures Example");

    //kill the process when the JFrame is closed
    setDefaultCloseOperation(JFrame.EXIT_ON_CLOSE);

    //create our FirstPersonView which serves two purposes
    // 1) it is our GLEventListener, and
    // 2) it is our KeyListener
    TexturesView lv = new TexturesView();

    //only three JOGL lines of code ... and here they are
    GLCapabilities glcaps = new GLCapabilities();
```

```
    glcanvas =
       GLDrawableFactory.getFactory().createGLCanvas(
glcaps);
    glcanvas.addGLEventListener(lv);

    //add the GLCanvas just like we would any
Component
    getContentPane().add(glcanvas,
BorderLayout.CENTER);
    setSize(500, 300);

    //center the JFrame on the screen
    centerWindow(this);
  }

  public void centerWindow(Component frame) {
    Dimension screenSize =
       Toolkit.getDefaultToolkit().getScreenSize();
    Dimension frameSize  = frame.getSize();

    if (frameSize.width  > screenSize.width )
       frameSize.width  = screenSize.width;
    if (frameSize.height > screenSize.height)
       frameSize.height = screenSize.height;

    frame.setLocation (
      (screenSize.width  - frameSize.width ) >> 1,
      (screenSize.height - frameSize.height) >> 1
    );
  }
}
import java.io.*;
import java.nio.*;
import javax.imageio.*;
import java.awt.image.*;

import java.awt.*;
import net.java.games.jogl.*;
import net.java.games.jogl.util.*;

/**
 * For this program only one of the
 * GLEventListener's methods matter. It
 * would be display().
```

Gene Davis

```
 *
 * It would be good to move the loading
 * and preparing of the texture to the init
 * method, but this code will be a bit
 * tricky to follow the first time, so I
 * will leave it mostly in the display()
 * method.
 */
public class TexturesView implements GLEventListener {

  public void init(GLDrawable drawable) {}

  /**
   * Take care of drawing here.
   */
  public void display(GLDrawable drawable) {
    GL gl = drawable.getGL();
    GLU glu = drawable.getGLU();
    GLUT glut = new GLUT();

    // We're changing the mode to GL.GL_PROJECTION
    // This is where we set up the camera
    gl.glMatrixMode(GL.GL_PROJECTION);
    gl.glLoadIdentity();

    // Aspect is width/height
    double w = ((Component) drawable).getWidth();
    double h = ((Component) drawable).getHeight();
    double aspect = w/h;
    // Remember, when using gluPerspective near
    // and far need to be positive.
    // The arguments are:
    // fovy, aspect, near, far
    glu.gluPerspective(60.0, aspect, 2.0, 10.0);

    gl.glMatrixMode(GL.GL_MODELVIEW);
    gl.glLoadIdentity();

    gl.glClearColor(0.0f, 0.0f, 0.0f, 1.0f);

    //Define points for 'eye', 'at' and 'up'
    //This is your camera. It ALWAYS goes
    //in the GL_MODELVIEW matrix.
    glu.gluLookAt(
                   0, 0, 0,
```

```
            0, 0, 20,
            0, 1, 0
         );

   // Time to set up the light. We will only
   // use one light. If we used multiple lights,
   // we would want to dim them so as to not
   // bleach out the scene with too much light.
   float position[] = {0f, 15f, -30f, 0f};
   gl.glLightfv(GL.GL_LIGHT0, GL.GL_POSITION,
position);

   float diffuse[] = {.7f, .7f, .7f, 0f};
   gl.glLightfv(GL.GL_LIGHT0, GL.GL_DIFFUSE,
diffuse);

   float ambient[] = {.2f, .2f, .2f, 0f};
   gl.glLightfv(GL.GL_LIGHT0, GL.GL_AMBIENT,
ambient);

   gl.glEnable(GL.GL_LIGHTING);
   gl.glEnable(GL.GL_LIGHT0);
   // GL_DEPTH_TEST prevents us from seeing polygons
   // that should be obscured. Remember to clear the
   // GL.GL_DEPTH_BUFFER_BIT
   gl.glEnable(GL.GL_DEPTH_TEST);

   // Notice the depth buffer bit is also being
cleared
   gl.glClear(
      GL.GL_COLOR_BUFFER_BIT |
      GL.GL_DEPTH_BUFFER_BIT
   );

   // Drawing a brick wall behind the scene:
   // First we set up a material to mix with
   // the texture. White is mostly harmless.
   float texMatMix[] = {1.0f, 1.0f, 1.0f, 1.0f};
   gl.glMaterialfv(
      GL.GL_FRONT_AND_BACK,
      GL.GL_AMBIENT_AND_DIFFUSE,
      texMatMix
   );

   gl.glEnable(GL.GL_TEXTURE_2D);
```

```
int wallTexture;
int retval[] = new int[1];

// We need to generate a number
// to associate texture with
// and place it into 'wallTexture'
gl.glGenTextures(1, retval);
wallTexture = retval[0];

// select the texture to work with
gl.glBindTexture(GL.GL_TEXTURE_2D, wallTexture);

// I have included this method to
// load an image from our jar and
// call the gluBuild2DMipmaps()
mipmapsFromPNG(glu, "an_img.png");

// Tell JOGL how to place the
// texture on our object

// I used the x and y magnification to fix the
// image distortion. The bricks will look more
// like they did in the original photo than in
// the final texture

// These parameters are:
// x magnification, x shear,
// z magnification, x translation
float sMapping[] = { 0.1f, 0f, 0f, 0f };
// y shear, y magnification,
// z magnification, y translation
float tMapping[] = { 0f, 0.2f, 0f, 0f };
// z mag. above is not useful to us, but the other
// parameters are worth playing with

// here we do lots of magic to make the sMapping
// and tMapping take affect
gl.glTexGeni(
    GL.GL_S,
    GL.GL_TEXTURE_GEN_MODE,
    GL.GL_OBJECT_LINEAR
);
gl.glTexGenfv(GL.GL_S, GL.GL_OBJECT_PLANE, sMapping);
```

```
    gl.glTexGeni(
        GL.GL_T,
        GL.GL_TEXTURE_GEN_MODE,
        GL.GL_OBJECT_LINEAR
    );
    gl.glTexGenfv(GL.GL_T, GL.GL_OBJECT_PLANE,
tMapping);

    gl.glEnable(GL.GL_TEXTURE_GEN_S);
    gl.glEnable(GL.GL_TEXTURE_GEN_T);


    // the object to draw the texture on is a quad
    gl.glBegin(GL.GL_QUADS);
    gl.glVertex3i(7, -3, 8);
    gl.glVertex3i(7, 3, 8);
    gl.glVertex3i(-7, 3, 8);
    gl.glVertex3i(-7, -3, 8);
    gl.glEnd();

    // stop using textures
    gl.glDisable(GL.GL_TEXTURE_2D);


    // Here we draw the objects. Notice
    // the use of glPushMatrix() and
    // glPopMatrix().

    // Placing icosahedron
    // The icosahedron takes a GL instance
    // as a parameter.
    gl.glPushMatrix();
    gl.glTranslated(0, 0, 5);
    float material1[] = {0.9f, 0.6f, 0.06f, 1.0f};
    gl.glMaterialfv(
        GL.GL_FRONT_AND_BACK,
        GL.GL_AMBIENT_AND_DIFFUSE,
        material1
    );
    glut.glutSolidIcosahedron(gl);
    gl.glPopMatrix();

    // Placing cube
    // The cube takes a GL instance and
```

```
    // a edge length as arguments.
    gl.glPushMatrix();
    gl.glTranslated(3, 0, 5);
    float material2[] = {0.9f, 0f, 0f, 1.0f};
    gl.glMaterialfv(
       GL.GL_FRONT_AND_BACK,
       GL.GL_AMBIENT_AND_DIFFUSE,
       material2
    );
    glut.glutSolidCube(gl, 1.2f);
    gl.glPopMatrix();

    //   Placing sphere
    //   The Cone parameters are:
    //      GLU instance (NOT GL)
    //      radius of base
    //      height of cone
    //      slices or subdivisions around cone
    //      stacks or subdivisions from base to peak
    gl.glPushMatrix();
    gl.glTranslated(-3, 0, 5);
    float material3[] = {0.05f, 0f, 0.7f, 1.0f};
    gl.glMaterialfv(
       GL.GL_FRONT_AND_BACK,
       GL.GL_AMBIENT_AND_DIFFUSE,
       material3
    );
    glut.glutSolidSphere(glu, .8, 100, 360);
    gl.glPopMatrix();

}

public void reshape(
                 GLDrawable drawable,
                 int x,
                 int y,
                 int width,
                 int height
              ) {}

public void displayChanged(
                    GLDrawable drawable,
                    boolean modeChanged,
                    boolean deviceChanged
                 ) {}
```

```java
/**
 * Assumes an RGB PNG with 8 bits per channel.
 * It may work with some other image types though.
 */
public void mipmapsFromPNG(GLU glu, String imgName)
{
    try {

        // load 'an_img.png' from the
        // JAR file as an InputStream
        InputStream is =
            ClassLoader.getSystemResourceAsStream(imgName);

        // buffer stream for better performance
        BufferedInputStream bis = new BufferedInputStream( is );

        // create a buffered image
        BufferedImage bi = ImageIO.read( bis );

        // make byte array
        Raster r = bi.getRaster();
        DataBufferByte dbi = (DataBufferByte) r.getDataBuffer();
        byte b[] = dbi.getData();

        // create the final byte buffer
        // to use as our texture
        ByteBuffer texture = null;
        int texSize = b.length;
        texture = ByteBuffer.allocateDirect( texSize );
        ByteOrder newOrder = ByteOrder.nativeOrder();
        texture.order(newOrder);

        // finally filling 'texture' byte buffer
        // with the texture found in byte array 'b'
        texture.put(b, 0, texSize);

        // the texture width is bi.getWidth()
        // the texture height is bi.getHeight()
        // the actual texture bytes are 'texture'
        int imgWidth = bi.getWidth();
        int imgHeight = bi.getHeight();
```

```
        // gluBuild2DMipmaps takes parameters for:
        //
        //   GL.GL_TEXTURE_2D - specifies a 2 dimensional
        //                     texture will be used
        //
        //   GL.GL_RGB8 - specifies the internal format
        //                of the image data
        //
        //   imgWidth - width of image (must be power of 2)
        //
        //   imgHeight - height of image (must be power of 2)
        //
        //   GL.GL_RGB - type of texels (pixels)
        //
        //   GL.GL_UNSIGNED_BYTE - format of texels (pixels)
        //
        //   texture - the array of texel info

        glu.gluBuild2DMipmaps(
                    GL.GL_TEXTURE_2D,
                    GL.GL_RGB8,
                    imgWidth,
                    imgHeight,
                    GL.GL_RGB,
                    GL.GL_UNSIGNED_BYTE,
                    texture
                    );

    } catch (Exception e) {
      System.out.println("Oops!");
      e.printStackTrace();
    }
  }
}
```

In Conclusion

That's all folks. You have enough tools at your disposal now to make some pretty cool applications.

Every topic mentioned in this book was just touched on lightly. I recommend browsing the internet for more information. At the time of my writing this book, there are no other books or manuals for learning JOGL. Hopefully this changes.

In the meantime, OpenGL books and tutorials may be a great source of information. Like I said before, C looks just like Java if you squint really hard.

Appendix A
JOGL Online Resources

Various Web Resources

The JOGL Project
https://jogl.dev.java.net/

"Jogl - User's Guide"
https://jogl.dev.java.net/nonav/source/browse/*checkout*/jogl/doc/userguide/index.html?rev=HEAD&content-type=text/html

Precompiled Binaries
https://games-binaries.dev.java.net/build/index.html

JOGL Demos
https://jogl-demos.dev.java.net/

"Jumping into JOGL" by Chris Adamson 09/11/2003
http://today.java.net/pub/a/today/2003/09/11/jogl2d.html

"Mixing Heavy and Light Components" by Amy Fowler. This one will give you an understanding of how to use GLCanvas with Swing.
http://java.sun.com/products/jfc/tsc/articles/mixing/

"How to: Getting started with JOGL". This is a really good thread on JOGL setup.
http://www.javagaming.org/cgi-bin/JGNetForums/YaBB.cgi?board=jogl;action=display;num=1058027992;start=

"From the Trenches". This appears to be a blog refering to JOGL development. Interesting read especially for OS X users.
http://www.jroller.com/page/gregorypierce/20031021

"gerard ziemski project page". This has some more info specific to OS X. It also has interesting comments on competing projects like gl4java.
http://homepage.mac.com/gziemski/projects/

"jogl - Java Bindings for OpenGL" Moderators: ChrisM, shawnkendall. This is a JOGL specific forum. Good for finding help or helping out others.
http://www.javagaming.org/cgi-bin/JGNetForums/YaBB.cgi?board=jogl

"beginner cannot run". A thread on the Jogl Forum.
http://www.javagaming.org/cgi-bin/JGNetForums/YaBB.cgi?board=jogl;action=display;num=1060985583;start=0#0

"3D graphics programming in Java, Part 3: OpenGL" by Bill Day of Java World's site.
http://www.javaworld.com/javaworld/jw-05-1999/jw-05-media_p.html

"The Puppy Games forum". Excellent thread reply by Chris Kline. A must read for JOGL background information.
http://www.puppygames.net/forums/viewtopic.php?t=38&start=30

Appendix B
OpenGL Online Resources

Various Web Resources

These links are all to C or C++ tutorials on OpenGL, but if you squint really hard it will look just like Java. So don't let any C like gibberish put you out.

OpenGL Home Page: The Industry's Foundation for High Performance Graphics
http://www.opengl.org/

OpenGL Survival Kit Tutorial by Nicole Deflaux Terry
http://www.cs.tulane.edu/www/Terry/OpenGL/Introduction.html

"Game Tutorials: Game Programming with Personality, From Start to Finish"
http://www.gametutorials.com/Tutorials/OpenGL/OpenGL_Pg1.htm

Light House 3D: OpenGL Tutorials
http://www.lighthouse3d.com/opengl/tutorials.shtml

A Quick OpenGL Tutorial by Will Weisser
http://www.erc.msstate.edu/~wyh/opengl/OpenGL_quick_tour.html

Help stamp out GL_PROJECTION abuse
http://www.sjbaker.org/steve/omniv/projection_abuse.html

OpenGL Programming Guide
http://www.esil.univ-mrs.fr/~smavroma/Docs/redbook/

OpenGL Correctness Tips
http://www.newcyber3d.com/asia/selfstudy/tips/ogl_correctness%20Tips.htm

Appendix C
Matrix Math

Matrices Are Cool

Let's face it, before a certain set of movies came along, no one knew what a matrix was. Since that famous question ("What is the Matrix?") was asked, pretty much no one still knows what a matrix is.

It's amazing what Hollywood can do for society, isn't it?

A matrix looks a lot like that screen saver they always had running on the computers in the Matrix movies. Typically it looks something like this:

$$\begin{bmatrix} 12 & 5 & -10 \\ 3 & 14 & 0 \\ 34 & 13 & 4 \\ -5 & -1 & -13 \end{bmatrix}$$

or,

$$\begin{bmatrix} 1 & 0 & 0 \\ 0 & 1 & 0 \\ 0 & 0 & 1 \end{bmatrix}$$

They come in all kinds of shapes and sizes. Any matrix that comes in a square n x n configuration with all zeroes except a diagonal line of ones like the matrix above is called an identity matrix. That's a special type of matrix that we will examine later.

Now before we get too far I'll warn you, this chapter is not an exhaustive treatise on matrices. For that, you should track down some good math texts. Some of what I describe will be simplified enough to make Arthur Cayley groan and roll over in his grave. The goal is to make matrix math approachable for the first timer.

Two Dimensional Arrays

Matrices are two dimensional arrays. Java doesn't have true two dimensional arrays. Usually two dimensional arrays are displayed as an array of arrays. For most purposes this works fine.

```
//here is a 3x3 array of integers
//this simulates a two dimensional
//array
int i[][] = new int[3][3];
```

OpenGL, which JOGL binds to via JNI, uses real two dimensional arrays. You see, C (chuckle) uses real two dimensional arrays. In C, a two dimensional array can be accessed as though it is one array by using pointers. This can't be done in Java. When giving C a two dimensional array from Java you have two choices. One, you can give it multiple arrays. This usually won't work. So, instead create one really long array from the multiple arrays, by tacking on array after array and pass this to C. This will work. Another option is to just use one array in Java and that way you won't have to do any conversions.

But What Are They?

Matrices are ways to describe change. The change may be to a position, or to a color. Moving objects or the viewers perspective in JOGL is called a transformation.

When you get ready for the nitty-gritty of matrices, you will want to look at the related topics of "Systems of Equations" and Determinants.

The uses of matrix math goes well into the realms of Business and Physics, as well as 3D graphics and many other fields.

Now suppose that you have three equations that are all related, such as

$$4x - y + 3z = -10$$
$$x - 2y + 2z = 9$$
$$6x - y + 5z = 2$$

you could convert this into a matrix that looks like this:

$$\begin{bmatrix} 4 & -1 & 3 & -10 \\ 1 & -2 & 2 & 9 \\ 6 & -1 & 5 & 2 \end{bmatrix}$$

Each equation above has an x, y and z variable. They also all have a result. All we did to create a matrix from the equations was line up the like variables and then remove the variables and the equals symbol and slap them inside a large set of square braces.

See how much fun making a matrix is?

Fun With Matrices

Among the fun things you can do with matrices, are subtraction, addition, division and multiplication. Here are the rules for multiplication and matrices.

When it comes to matrix multiplication AB != BA. They are not the same. If you want to multiply two matrices to obtain a new one, the first matrix MUST have the same number of columns as the second matrix has rows. This means that not all matrices may be multiplied with each other.

Here is a simple matrix multiplication. I'm not going to explain detail. It just doesn't lend itself to English. Look to the example for your explanation.

$$\begin{bmatrix} a & b & c \end{bmatrix} \begin{bmatrix} d \\ e \\ f \end{bmatrix} = \begin{bmatrix} ad + be + cf \end{bmatrix}$$

Multiplying these two matrices results in a matrix of one row and one column. If this looks tedious, think of handling a complex matrix multiplication by hand. It makes you appreciate computers real fast.

The idea is that the number of rows in the first matrix will be the number of rows in the solution. The number of columns in the second matrix will be the number of columns in the final solution.

Now let's do a more complex matrix multiplication so that you can see how it is done.

$$\begin{bmatrix} 2 & -1 \\ 3 & 0 \end{bmatrix} \begin{bmatrix} -4 & 8 \\ 5 & 7 \end{bmatrix} = ?$$

$$2(-4) - 1(5) = \begin{bmatrix} -13 & ? \\ ? & ? \end{bmatrix}$$

$$2(8) - 1(7) = \begin{bmatrix} -13 & 9 \\ ? & ? \end{bmatrix}$$

$$3(-4) - 0(5) = \begin{bmatrix} -13 & 9 \\ -12 & ? \end{bmatrix}$$

$$3(8) - 0(7) = \begin{bmatrix} -13 & 9 \\ -12 & 24 \end{bmatrix}$$

Who Are You?

Now let's look at what is so special about the identity matrices. Yes there are more than one. Here are a few examples of identity matrices:

$$\begin{bmatrix} 1 & 0 \\ 0 & 1 \end{bmatrix}$$

$$\begin{bmatrix} 1 & 0 & 0 \\ 0 & 1 & 0 \\ 0 & 0 & 1 \end{bmatrix}$$

$$\begin{bmatrix} 1 & 0 & 0 & 0 \\ 0 & 1 & 0 & 0 \\ 0 & 0 & 1 & 0 \\ 0 & 0 & 0 & 1 \end{bmatrix}$$

See any patterns emerging? Identity matrices always have as many columns as rows. Also they have a diagonal line of ones in a sea of zeros. I'm sure there are some philosophical thoughts that could be explored here. We'll skip that.

Now multiply the following using the rules previously discussed.

$$\begin{bmatrix} 2 & 5 & -4 \\ 3 & -1 & 0 \\ 9 & 6 & -4 \end{bmatrix} \begin{bmatrix} 1 & 0 & 0 \\ 0 & 1 & 0 \\ 0 & 0 & 1 \end{bmatrix}$$

You'll notice that the result is that nothing happens. You end up with the same matrix that you started with. The identity matrix is the "multiplicative identity matrix." Try saying that ten times fast!

The identity matrix is used for initializing matrix modes in JOGL. This is because the first transformation that happens to the mode will result in nothing but the transformation being inserted unchanged.

Index

A

Acknowledgments x
ActionEvent 33, 41
ActionListener 31, 33, 38, 39, 41
actionPerformed 33, 41
addActionListener 32, 40
addEventListener 7
addFocusListener 73
addGLEventListener 10, 11, 15, 24, 32, 40, 49, 58, 65, 73, 78, 85, 90, 103, 114, 120, 131, 143
addKeyListener 78, 82, 120
addMouseListener 85
addMouseMotionListener 90
addSuchAndSuchListener 5
ambient 127, 128, 133, 145
angle 46, 47, 51, 66, 67, 68, 117
angles 46
animate 55, 56
animation 52, 56, 63
Animator 55, 56, 57, 58, 63, 64, 65
API 1, 2
automate 137
AWT 3, 5, 6, 7, 11, 55, 70, 77, 108
AWTEvent 72
axis 19, 20, 27, 28, 37, 45, 59, 88, 93

B

background 129
binaries 2, 3, 153
binary 2, 3
binding 8
bindings 2
bitmaps 107
BorderLayout 10, 15, 24, 32, 41, 49, 58, 65, 73, 79, 85, 90, 103, 114, 120, 131, 143
BouncingDisplay 58, 59
bricks 138, 146
brightness 127
buffer 129, 133, 145, 149
buffered 149
BufferedImage 149
BufferedInputStream 149
bug 38, 56
build 2, 3, 153
ButtonGroup 40
ByteBuffer 149
ByteOrder 149

C

camera 97, 98, 99, 100, 101, 102, 104, 112, 115, 118, 119, 121, 122, 129, 132, 133, 137, 144
Canvas 22, 23, 70
Cartesian 19, 21, 23, 24, 26, 36, 44
ceiling 127
center 10, 15, 24, 32, 41, 47, 49, 50, 58, 65, 68, 73, 79, 85, 90, 101, 103, 114, 120, 131, 138, 143
cgi 154
circle 46, 47, 48, 49, 50, 51, 63, 68
Class 141
ClassLoader 149
classpath 2, 9
click 5, 71
clock 53, 54
clone 30, 138, 139
color 21, 22, 23, 30, 34, 38, 60, 66, 72, 74, 76, 126, 127,

128, 129, 137, 159
compile 9, 38, 142
Component 3, 10, 15, 24, 25, 32, 41, 49, 58, 65, 71, 73, 77, 79, 82, 85, 88, 90, 93, 103, 104, 114, 115, 120, 122, 131, 132, 143, 144
ComponentEvents 84
concave 110
cone 134, 148
configuration 158
converge 98
conversion 5, 137
convex 110
coordinate 19, 20, 21, 23, 29, 60, 62, 66, 68, 80, 83, 84, 87, 88, 89, 92, 93, 96, 100, 110, 111, 112, 118
coordinates 21, 47, 48, 95, 100, 101, 128
cosine 38, 40, 43, 47, 50, 51, 67
cross 1
cube 134, 147
curve viii
cut 21, 112
cylinder 111, 112

D

DataBufferByte 149
debugging 140
decimal 4, 5
define 47, 63, 99, 100, 101
degree 47, 68
demos 153
depth 12, 81, 88, 93, 98, 126, 129, 133, 145
descendant 72
design 55, 99, 110
destroy 111
diagonal 18, 158, 161
diagram 46

dial 46, 63
DialDisplay 64, 65
diffuse 127, 128, 133, 145
digital 137
dimension 96
disabled 127
discouraged viii
disk 111, 112
displays 22, 38
disposal 150
distortion 146
distribution 3
divide 22
dll 9
doc 56, 153
dodecahedrons 112
dot 108
double 4, 34, 35, 47, 50, 51, 66, 67, 68, 88, 93, 96, 100, 101, 104, 109, 111, 115, 117, 122, 132, 141, 144
dragging 89
draw 5, 11, 16, 23, 26, 27, 28, 29, 30, 35, 36, 37, 38, 43, 44, 45, 47, 62, 110, 111, 118, 140, 147
drawing 7, 11, 12, 16, 21, 22, 23, 26, 27, 29, 34, 36, 42, 43, 44, 50, 60, 68, 75, 80, 84, 87, 92, 98, 104, 107, 108, 111, 112, 115, 118, 119, 122, 132, 144
driven 70

E

edge 46, 47, 63, 111, 134, 140, 148
effect 100, 140
efficiency 56
element 128, 129
email vi

equation 30, 159
erase 21, 60
error 4
Event 5, 70, 71
EventListener 71
EventObject 71, 72
events 1, 5, 38, 71
Exception 9, 55, 150
exit 56
EXIT_ON_CLOSE 10, 15, 24, 31, 39, 48, 58, 64, 73, 78, 85, 90, 102, 114, 120, 131, 142
extending 6, 55
extends 6, 9, 14, 24, 31, 39, 48, 54, 57, 64, 70, 72, 78, 84, 89, 102, 113, 119, 130, 142
eye 48, 101, 104, 110, 115, 122, 133, 144
eyeX 101
eyeY 101
eyeZ 101

F

face 1, 157
facing 117
far 7, 52, 55, 56, 59, 98, 99, 100, 101, 102, 104, 105, 107, 108, 115, 117, 122, 126, 132, 141, 144, 158
farther 98, 100
field 100
fields viii, 159
FirstCircle 48
FirstCircleEventListener 49, 50
FirstPersonMovement 119
FirstPersonView 118, 120, 121, 131, 142
float 4, 16, 23, 27, 34, 43, 44, 50, 59, 60, 66, 67, 74, 80, 86, 91, 104, 115, 121, 127, 128, 129, 133, 134, 140, 145, 146, 147, 148
floats 4, 5, 34, 59, 118
focus 5, 11, 72, 76, 77, 82, 117, 139
FocusEvent 75, 76, 77
FocusEvents 76
FocusExample 72, 76
FocusExampleDisplay 73, 74, 76
focusGained 75, 77
FocusListener 74, 75, 76, 77, 79
FocusListeners 72
focusLost 76, 77
fonts 107
format 96, 139, 150
formula 47, 48, 61
fovy 100, 105, 115, 122, 132, 144
frame 10, 11, 15, 16, 25, 32, 33, 41, 49, 58, 61, 65, 73, 74, 79, 85, 86, 90, 91, 103, 114, 120, 131, 143
frustum 99
further 136

G

gaming 1
genedavissoftware vi, ix, x, 141
geometric 110, 111
geometry 20
getActionCommand 41
getComponent 77, 84, 88, 93, 118
getContentPane 10, 15, 24, 32, 40, 41, 49, 58, 65, 73, 79, 85, 90, 103, 114, 120, 131, 143
getData 149
getDataBuffer 149
getDefaultToolkit 10, 15, 25, 32, 41, 49, 58, 65, 74, 79, 85, 90, 103, 114, 120, 131, 143
getFactory 10, 11, 15, 24, 32, 40, 49, 58, 64, 73, 78, 85, 90, 103, 114, 120, 131, 143

getGL 7, 16, 26, 34, 42, 50, 51, 59, 60, 66, 67, 75, 80, 86, 87, 91, 92, 103, 104, 115, 121, 122, 132, 144
getGLU 7, 16, 26, 34, 42, 50, 59, 66, 75, 80, 86, 91, 103, 104, 115, 121, 122, 132, 144
getHeight 88, 93, 104, 115, 122, 132, 144, 149
getKeyChar 81, 82, 83, 124
getKeyCode 83
getKeyText 83
getRaster 149
getScreenSize 10, 15, 25, 32, 41, 49, 58, 65, 74, 79, 85, 90, 103, 114, 120, 131, 143
getSize 10, 15, 25, 32, 41, 49, 58, 65, 74, 79, 85, 90, 103, 114, 120, 131, 143
getSystemResourceAsStream 149
getText 33
getWidth 88, 93, 104, 115, 122, 132, 144, 149
getX 83, 88, 93
getY 83, 88, 93
Gimp 137
GL 1, 3, 7, 8, 16, 17, 21, 23, 26, 27, 28, 29, 30, 34, 35, 36, 37, 38, 42, 43, 44, 45, 47, 50, 51, 59, 60, 62, 66, 67, 68, 75, 80, 81, 86, 87, 91, 92, 96, 97, 103, 104, 105, 107, 108, 109, 110, 112, 115, 116, 117, 121, 122, 126, 127, 128, 129, 132, 133, 134, 141, 144, 145, 146, 147, 148, 150
GL_AMBIENT 128, 129, 133, 134, 145, 147, 148
GL_AMBIENT_AND_DIFFUSE 129, 134, 145, 147, 148
GL_COLOR_BUFFER_BIT 17, 21, 23, 27, 36, 44, 51, 60, 67, 75, 80, 87, 92, 105, 116, 122, 133, 145
GL_DEPTH_BUFFER_BIT 129, 133, 145
GL_DEPTH_TEST 129, 133, 145
GL_DIFFUSE 127, 133, 145
GL_FRONT_AND_BACK 129, 134, 145, 147, 148
GL_LIGHT0 127, 128, 133, 145
GL_LIGHT1 127
GL_LIGHT2 127
GL_LIGHT3 127
GL_LIGHT5 128
GL_LIGHT7 128
GL_LIGHTING 126, 133, 145
GL_LINES 27, 30, 37, 44, 45
GL_MAX_LIGHTS 127
GL_MODELVIEW 96, 97, 103, 104, 105, 115, 116, 117, 121, 122, 132, 133, 141, 144
GL_OBJECT_LINEAR 146, 147
GL_OBJECT_PLANE 146, 147
GL_POINTS 17, 35, 36, 38, 43, 62, 81, 87, 92
GL_POLYGON 47, 51, 67, 68, 110
GL_POSITION 128, 133, 145
GL_PROJECTION 16, 26, 34, 42, 50, 60, 66, 75, 80, 87, 92, 96, 97, 104, 115, 121, 132, 144
GL_QUADS 105, 147
GL_RGB 150
GL_RGB8 150
GL_S 146
GL_T 28, 37, 45, 145, 146, 147, 150
GL_TEXTURE_2D 145, 146, 147, 150

GL_TEXTURE_GEN_MODE 146, 147
GL_TEXTURE_GEN_S 147
GL_TEXTURE_GEN_T 147
GL_TRIANGLES 28, 37, 45
GL_UNSIGNED_BYTE 150
gl4java 2, 154
glBegin 17, 27, 28, 30, 35, 36, 37, 38, 43, 44, 45, 47, 51, 62, 67, 68, 81, 87, 92, 105, 108, 110, 111, 147
glBindTexture 146
GLCanvas 3, 7, 10, 11, 12, 13, 15, 21, 22, 23, 24, 29, 31, 32, 33, 38, 39, 40, 41, 48, 49, 50, 55, 56, 57, 58, 59, 60, 63, 64, 65, 66, 68, 70, 71, 72, 73, 74, 76, 77, 78, 80, 81, 82, 84, 85, 86, 87, 88, 89, 90, 91, 92, 93, 103, 114, 118, 119, 120, 121, 128, 130, 131, 142, 143
GLCanvases 7, 11, 56, 72, 73, 76, 77
GLCapabilities 3, 9, 10, 11, 15, 24, 29, 32, 40, 49, 58, 64, 73, 78, 85, 90, 102, 114, 120, 131, 142
GLCavas 89
GLClass 82
glClear 17, 21, 23, 27, 36, 44, 51, 60, 67, 75, 80, 87, 92, 105, 116, 122, 129, 133, 145
glClearColor 16, 21, 22, 26, 29, 34, 42, 50, 60, 66, 75, 80, 87, 92, 104, 115, 122, 133, 144
glColor3f 17, 23, 27, 35, 36, 37, 43, 44, 45, 51, 60, 67, 81, 87, 92, 105, 116, 122, 123, 126

gldFactory 58, 64, 65
glDisable 147
GLDrawable 3, 7, 11, 12, 13, 16, 17, 26, 34, 36, 42, 43, 44, 50, 51, 56, 59, 60, 62, 66, 68, 75, 80, 81, 86, 87, 88, 91, 92, 93, 103, 104, 105, 106, 115, 116, 121, 122, 124, 132, 134, 135, 144, 148
GLDrawableFactory 3, 10, 11, 15, 24, 32, 40, 49, 58, 64, 73, 78, 85, 90, 103, 114, 120, 131, 143
GLDrawables 7, 13
glEnable 126, 127, 129, 133, 145, 147
glEnd 17, 27, 28, 30, 35, 36, 37, 38, 43, 45, 46, 48, 51, 62, 67, 68, 81, 87, 92, 105, 108, 110, 111, 147
GLEventListener 5, 11, 12, 13, 16, 18, 25, 29, 33, 34, 41, 42, 49, 50, 58, 59, 64, 65, 74, 76, 78, 79, 82, 85, 86, 89, 90, 91, 103, 114, 118, 119, 120, 121, 131, 132, 142, 143, 144
GLEventListeners 5, 7, 12, 16, 25, 29, 33, 42, 79, 86, 91, 121
glFrustrum 97
glFrustum 97, 104
glGenTextures 146
GLJPanel 3, 7, 12, 70, 71, 77, 81, 82, 87, 92
GLJPanels 7, 11
glLightfv 127, 128, 133, 145
glLineWidth 26, 29, 34, 42, 66
glLoadIdentity 16, 26, 34, 42, 50, 60, 66, 75, 80, 87, 92, 97, 104, 105, 115, 121, 122, 132, 144

glLoadMatrix 97, 104
glMaterialfv 129, 134, 145, 147, 148
glMatrixMode 16, 26, 34, 42, 50, 60, 66, 75, 80, 87, 92, 97, 104, 105, 115, 121, 122, 132, 144
glOrtho 97, 104
glPointSize 17, 34, 36, 38, 42, 60, 80, 87, 92
glPopMatrix 140, 147, 148
glPushMatrix 140, 147, 148
glRotated 117
glScaled 117
glTexGenfv 146, 147
glTexGeni 146, 147
glTranslate 117
glTranslated 117, 118, 123, 124, 134, 141, 147, 148
glTranslatef 117, 118
GLU 7, 16, 26, 29, 34, 42, 50, 59, 66, 75, 80, 86, 91, 103, 104, 107, 110, 111, 112, 115, 116, 121, 122, 132, 134, 139, 144, 148, 149
glu 16, 26, 34, 42, 50, 59, 60, 66, 75, 80, 86, 87, 91, 92, 103, 104, 105, 107, 108, 111, 115, 116, 121, 122, 132, 133, 134, 139, 144, 146, 148, 149, 150
GLU_FILL 111
GLU_LINE 111, 116
GLU_POINT 111
GLU_SILHOUETTE 111
gluBuild2DMipmaps 146, 150
gluCylinder 111
gluDeleteQuadric 111, 116
gluDisk 111
GlueGen 8
gluLookAt 99, 100, 101, 102, 104, 116, 118, 122, 133, 144
gluNewQuadric 111, 116
gluOrtho2D 16, 21, 26, 29, 34, 42, 50, 60, 66, 75, 80, 87, 92, 97, 98, 104
gluPerspective 97, 98, 99, 100, 101, 102, 104, 105, 115, 119, 122, 129, 132, 144
GLUquadric 111, 116
gluQuadricDrawStyle 111, 116
gluSphere 111, 116
GLUT 1, 3, 107, 108, 112, 114, 116, 122, 132, 144
glut 107, 116, 122, 123, 124, 132, 134, 144, 147, 148
GlutObjects 113
GlutObjectsView 114
glutSolidCube 134, 148
glutSolidIcosahedron 134, 147
glutSolidSphere 134, 148
glutWireIcosahedron 116, 123, 124
glVertex 108, 109, 110, 111
glVertex2d 27, 28, 30, 35, 36, 37, 38, 43, 44, 45, 46, 47, 51, 62, 67, 68, 109
glVertex2i 17, 81, 87, 92, 109
glVertex3d 109
glVertex3i 105, 109, 147
glViewport 16, 21, 26, 29, 34, 42, 50, 60, 66, 75, 80, 87, 92
graph 20, 23, 30, 39, 50
graphic 13, 22
group 111
GUI 71

H

header 8, 107
height 11, 12, 15, 17, 25, 26, 33, 36, 41, 44, 49, 51, 58, 59, 60, 62, 65, 66, 68, 74, 75,

79, 81, 85, 86, 87, 88, 90,
91, 93, 100, 103, 104, 105,
111, 114, 115, 116, 120, 122,
124, 131, 132, 134, 135,
143, 144, 148, 149, 150
HelloWorld 8
HotSpot 2
html 2, 3, 153
hue 128

I

icosahedron 133, 147
identity 96, 98, 158, 161, 162
illuminating 126, 128
illusion 98, 110
illustration 57, 63
image 98, 136, 137, 138, 139, 140,
141, 143, 146, 149, 150
ImageIO 149
implements 6, 12, 13, 16, 25, 31,
34, 38, 39, 42, 50, 55, 59,
65, 71, 74, 79, 86, 91, 103,
114, 121, 132, 144
importing 139
include 137
index 2, 3, 153
industry 1
initialize 105
inner 111
input x, 38, 63, 127
InputStream 149
inserted 162
insets 50, 51
installation 2, 3, 8
integers 158
ints 5

J

JAR 139, 140, 141, 149
javac 9
Javadoc 56

javadoc 97, 109
JButton 32
JFrame 9, 10, 13, 14, 15, 24, 31,
32, 39, 41, 48, 49, 51, 57,
58, 63, 64, 65, 72, 73, 78,
79, 84, 85, 89, 90, 102, 103,
113, 114, 119, 120, 130, 131,
142, 143
JGNetForums 154
JLabel 32
JLabels 32
JNI 7, 8, 107, 158
JOGL viii, ix, x, 1, 2, 3, 5, 7, 8,
9, 10, 11, 13, 14, 15, 17, 18,
21, 24, 29, 30, 31, 32, 39,
40, 47, 48, 52, 55, 69, 70,
72, 73, 77, 78, 84, 85, 89,
90, 95, 96, 99, 102, 104,
107, 108, 113, 114, 117, 119,
120, 127, 128, 129, 130,
131, 136, 137, 138, 139,
140, 142, 146, 151, 153,
158, 159, 162
JPanel 32, 39, 40, 70
JRadioButton 40
JRadioButtons 40
JTextField 31
JTextFields 32
JVM 7, 11, 55, 56

K

keyboard 71, 76, 77, 108, 124
KeyDisplay 78, 79, 82, 85, 90
KeyEvent 81, 82, 83, 118, 124
KeyEvents 77, 82, 83
KeyExample 78, 82
KeyListener 77, 78, 79, 81, 82,
85, 90, 118, 120, 121, 124,
131, 142
KeyListeners 77, 118
keyPressed 81, 82, 83, 124

keyReleased 81, 82, 83, 124
keys 82, 83, 118, 124
keystrokes 82, 83
keyTyped 81, 82, 83, 124

L

languages 4
length 134, 148, 149
lens 99, 100
libraries 3, 7, 8, 9, 11, 97, 107, 108
library 2, 7, 8, 9, 108, 116
light 6, 7, 99, 126, 127, 128, 133, 137, 145
lighting 116, 123, 126, 127, 129, 130
LightingApp 130, 131
LightingView 131, 132
lights 127, 128, 133, 145
LineGLEventListener 31, 34, 38
LineGraphApp 31, 38
listener 5, 31, 32, 33, 39, 40, 41, 70, 71, 76, 84, 89
loading 139, 144
longitude 111
luck 38, 77

M

machine 126
magnify 140
main 4, 6, 7, 8, 9, 14, 23, 24, 29, 31, 33, 39, 48, 53, 54, 55, 57, 64, 70, 72, 77, 78, 82, 84, 89, 96, 102, 107, 113, 119, 130, 136, 142
manifest 141
manual 56
material 129, 130, 145
math viii, 2, 4, 5, 22, 30, 69, 95, 96, 158, 159
matrix 69, 95, 96, 97, 98, 104, 116, 122, 133, 140, 141, 144, 157, 158, 159, 160, 162
milliseconds 54
mixing 6, 7, 22, 153
model 2, 5, 70, 71
modern 7
modes 97, 162
monitor 21, 29, 100
motion 89
Mouse 83, 89
mouse 5, 71, 72, 77, 83, 89, 108
mouseClicked 83, 84, 88
MouseDisplay 85, 86
mouseDragged 93
mouseEntered 83, 88
MouseEvent 88, 93
MouseEvents 83
MouseExample 84, 85
mouseExited 83, 88
MouseListener 83, 84, 86, 88
MouseListeners 83
MouseMotionDisplay 90, 91
MouseMotionExample 89, 90
MouseMotionListener 83, 89, 91, 93
MouseMotionListeners 89
mouseMoved 93
mousePressed 83, 88
mouseReleased 83, 88
movement 89, 118, 124
moving 117, 118
multiplying 63, 96
multithreaded 56

N

nature 99, 137
navigate 77, 123
near 98, 99, 100, 101, 102, 104, 105, 115, 122, 129, 132, 144
negative 20, 100
newbies 101

nightscape 99
nio 143
notify 71
nurbs 107
NVidia 56

O

Object 2, 21, 56
obvious 109
OpenGL viii, 1, 2, 3, 5, 7, 8, 11,
    13, 18, 19, 29, 69, 84, 89,
    95, 96, 107, 108, 110, 118,
    126, 127, 139, 151, 155, 158
OpentGL 2
origin 20, 50, 68, 101, 112, 117,
    118, 140, 141
orthographic 98
OurGraphApp 24, 28, 29
OurGraphGLEventListener 24,
    25, 29
oval 50, 51
overload 118
overwhelmed 18

P

paint 128
parallel 98
parameter 134, 140, 147
parent 12
parses 8
pattern 6, 33, 111, 137
perceive 126
percent 88, 93
performance 128, 149
perpendicular 19
perspective 98, 100, 105, 112, 159
PhotoShop 22, 137, 139
picture 14, 137
pie 46, 112
pixellated 138
pixels 35, 36, 60, 66, 80, 87, 92,
    138, 150
PlanesView 103
planet 99
platform 1
play 108, 130
PNG 137, 139, 149
png 141, 146, 149
point 13, 19, 20, 22, 30, 35, 36,
    38, 46, 57, 60, 62, 63, 80,
    84, 87, 89, 92, 99, 100, 101,
    108, 110, 111, 117, 136, 138
polygon 128
polymorphism 6, 7, 12
Pong 30
pong 57
pop 136, 140
positioning 118, 119, 140
potatoes 97
precision 4, 5
primitive 83, 112
primitives 4, 96, 110, 111
pyramid 99

Q

quad 110, 111, 116, 147
quadric 111
Quads 102
quads 102, 110

R

radian 47
radius 46, 47, 50, 51, 66, 67, 68,
    111, 134, 148
Raster 149
rectangle 21
reference 21, 117
render 99
repaint 33, 41, 55, 75, 76, 78, 82,
    84, 85, 88, 89, 90, 93, 118,
    120, 124
repetition 137

resources ix, 141
review viii, 3, 6, 39, 62, 70, 118
RGB 22, 23, 30, 128, 129, 137, 139, 149
rotate 117, 138, 139
row 160
Runnable 10, 14, 24, 31, 39, 48, 54, 55, 57, 64, 67, 72, 78, 84, 89, 102, 113, 119, 130, 142

S

scale 117, 127
screen 3, 10, 15, 21, 24, 30, 32, 41, 49, 58, 65, 73, 76, 77, 79, 85, 90, 95, 98, 103, 105, 114, 120, 131, 143, 157
SecondGLEventListener 15, 16, 18
SecondJoglApp x, 14, 15
setActionCommand 40
setDefaultCloseOperation 10, 15, 24, 31, 39, 48, 58, 64, 73, 78, 85, 90, 102, 114, 120, 131, 142
setGLCanvas 78, 80, 85, 86, 90, 91, 120, 121
setLayout 73
setLocation 11, 15, 25, 33, 41, 49, 58, 65, 73, 74, 79, 86, 91, 103, 114, 120, 131, 143
setResizable 73
setSize 10, 15, 24, 32, 41, 49, 58, 65, 73, 79, 85, 90, 103, 114, 120, 131, 143
setVisible 10, 14, 24, 31, 39, 48, 57, 64, 72, 78, 82, 84, 89, 102, 113, 119, 130, 142
SGI 1, 2
shading 126
shiftXPosition 47, 66, 67

shiftYPosition 47, 66, 67, 68
silhouette 126
SimpleGLEventListener 9, 10, 11, 12, 13
SimpleJoglApp 9, 10, 13
sine 38, 40, 42, 43, 47, 50, 51, 67
slices 111, 134, 148
slope 30, 32, 34, 35, 56, 59, 60, 61, 62, 63
solid 116, 123, 126
space 4, 118
sphere 111, 112, 116, 134, 148
stacks 111, 134, 148
sun 153
sunlight 126
Swing viii, 2, 3, 5, 6, 7, 11, 55, 70, 108
SwingUtilities 10, 14, 24, 31, 39, 48, 57, 64, 67, 72, 78, 84, 89, 102, 113, 119, 130, 142

T

tangent 38, 40, 43
TaskExample 54
template 9
TextArea 77
TextField 77
texture 137, 138, 139, 140, 141, 144, 145, 146, 147, 149, 150
textures 136, 147
TexturesApp 141, 142
TexturesView 142, 144
thickness 29
Thread 53, 54, 55, 56
ThreadExample 54
threading 53
time viii, 1, 4, 8, 19, 21, 23, 29, 53, 56, 57, 61, 68, 82, 97, 99, 102, 105, 107, 108, 110, 112, 118, 144, 151
Timer 53, 54, 55

timer 53, 158
TimerExample 53
TimerTask 53, 54
Toolkit 10, 15, 25, 32, 41, 49, 58, 65, 74, 79, 85, 90, 103, 108, 114, 120, 131, 143
toolkit 108
transformation 68, 159, 162
transformations 69, 117, 141
translate 117, 140
triangle 30
trig 35, 38, 43
trigger 5
TrigGLEventListener 39, 42
TrigGraphApp 39

U

UML 13
unit 62
universe 101
UnsatisfiedLinkError 9
up 2, 3, 5, 6, 7, 8, 9, 29, 32, 34, 35, 39, 40, 43, 47, 53, 54, 55, 56, 62, 63, 68, 69, 80, 87, 89, 92, 95, 96, 97, 99, 100, 101, 102, 104, 108, 111, 112, 115, 119, 121, 122, 126, 127, 128, 129, 132, 133, 138, 140, 144, 145, 159, 162

V

vertex 107
vertical 27, 37, 44, 139
vertices 108
ViewPort 103, 115, 121
viewport 60, 66
views 99
virtual 82, 83, 129
visible 63, 82, 126

W

website ix, 141
width 10, 11, 12, 15, 17, 25, 26, 29, 33, 36, 41, 44, 49, 51, 58, 60, 62, 65, 66, 68, 74, 75, 79, 81, 85, 86, 87, 88, 90, 91, 92, 93, 100, 103, 104, 105, 114, 115, 116, 120, 122, 124, 131, 132, 134, 138, 143, 144, 148, 149, 150
window 1, 29, 108, 124
words 5, 21, 47, 119

About The Author

Gene Davis has been programming professionally since 1996 with many years of Java experience. He currently works in the programming industry, writes fiction and programming tutorials, and runs two websites: www.genedavis.com and www.genedavissoftware.com. He resides in Woods Cross, Utah with his wife and children.